CONTENTS

Targeting Maths Homework Year 5

ISBN: 978 1 925726 58 9

Published by Pascal Press
PO Box 250
Glebe NSW 2037
www.pascalpress.com.au
contact@pascalpress.com.au

Author: Ann & Johnny Baker
Publisher: Lynn Dickinson
Editor: Ruth Schultz
Cover and Text Designer: Janice Bowles

Typesetter: Ruth Schultz
Images & Illustrations: Dreamstime (unless otherwise indicated)

Printed by Wai Man Bookbinding (China) Ltd

Introduction

Targeting Maths Homework aims to build and reinforce Maths skills. This book supports the ACARA Australian Version 9·0 Curriculum for Year 5 and helps children to revise and consolidate what has been taught in the classroom. ACARA codes are shown on each unit and a chart explaining their content descriptions is on pages v and vi. The inside cover shows the topics in each unit.

The structure of this book

This book has 32 carefully graded four-page units.
The units are divided between:

- ★ Number and Algebra
- ★ Measurement and Space
- ★ Statistics & Probability
- ★ Problem Solving.

Assessment

Term Reviews follow Units 1–8, 9–16, 17–24 and 25–32 to test work covered during the term, and allow parents and carers to monitor their child's progress. Children are encouraged to mark each unit as it is completed and to colour in the traffic lights at the end of each segment. These results are then transferred to the Marking Grid. Parents and carers can see at a glance if their child is excelling or struggling!

NAPLAN*-style questions

Each unit has a set of three NAPLAN*-style questions that directly match the learning in the unit and mirror question styles that frequently occur in NAPLAN. This is to ensure that children's familiarity and confidence with NAPLAN*-style questions is consistently developed.

Green = Excellent — 2 or fewer questions incorrect
Orange = Passing — 50% or more questions answered correctly
Red = Struggling — fewer than 50% correct and needs help

SCORE /10 0-4 5-8 9-10

How to Use This Book

The activities in this book are specifically designed to be used at home with minimal resources and support. Helpful explanations of key concepts and skills are provided throughout the book to help understand the tasks. Useful examples of how to do the activities are provided.

Regular practice of key concepts and skills will support the work your child does in school and will enable you to monitor their progress throughout the year. It is recommended that children complete 8 units per school term (one a week) and then the Term Review. Every unit has a Traffic Light scoreboard at the end of each section.

You or your child should mark each completed unit and then colour the traffic light that corresponds to the number of correct questions. This process will enable you to see at a glance how your child is progressing and to identify weak spots. The results should be recorded at the end of each term on the Marking Grid on page 1.

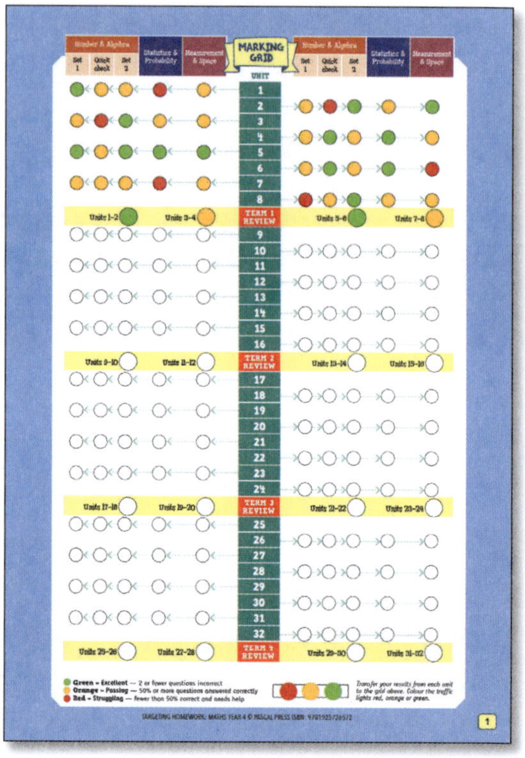

NOTE: The Problem of the Week sections do not appear on the Marking Grid as they often have multiple or subjective answers that cannot be easily scored.

The Term Review results are important for tracking progress and identifying any improvements in performance. If you find that certain questions are repeatedly causing difficulties and errors, then there is a good reason to discuss this with your child's teacher and arrange for extra instruction in that problem area.

Quick Maths Facts

The Quick Maths Facts on page 146 include multiplication tables up to 12 times, measurement equivalents, simple decimal–fraction–percentage conversions, a visual reminder of place value into the millions and the 'Days of the Month' rhyme. An accurate ruler has also been provided. Copy and laminate this page so that it is always available as a handy reference for your child.

Answers

The answer section on pages 147–164 can be removed, stapled together and kept somewhere safe. Use it to check answers when your child has completed each unit. Encourage your child to colour in the Traffic Light boxes when the answers have been calculated.

Australian Curriculum Correlations V9

Year 5 Maths		Number & Algebra	Statistics & Probability	Measurement & Space
ACARA CODE	**CONTENT DESCRIPTION**	**UNITS**	**UNITS**	**UNITS**
NUMBER				
AC9M5N01	Interpret, compare and order numbers with more than 2 decimal places, including numbers greater than one, using place value understanding; represent these on a number line	1, 9, 16, 17, 25		
AC9M5N02	Express natural numbers as products of their factors, recognise multiples and determine if one number is divisible by another	2, 9, 12, 17, 25, 29		
AC9M5N03	Compare and order fractions with the same and related denominators including mixed numerals, applying knowledge of factors and multiples; represent these fractions on a number line	3, 11, 16, 18, 24, 26, 31		
AC9M5N04	Recognise that 100% represents the complete whole and use percentages to describe, represent and compare relative size; connect familiar percentages to their decimal and fraction equivalents	4, 11, 18, 26, 31		
AC9M5N05	Solve problems involving addition and subtraction of fractions with the same or related denominators, using different strategies	3, 12, 19, 27		
AC9M5N06	Solve problems involving multiplication of larger numbers by one- or two-digit numbers, choosing efficient calculation strategies and using digital tools where appropriate; check the reasonableness of answers	5, 10, 20, 27, 28		
AC9M5N07	Solve problems involving division, choosing efficient strategies and using digital tools where appropriate; interpret any remainder according to the context and express results as a whole number, decimal or fraction	6, 8, 13, 21, 22, 26		
AC9M5N08	Check and explain the reasonableness of solutions to problems including financial contexts using estimation strategies appropriate to the context	7, 13, 14, 20, 21, 24, 29		
AC9M5N09	Use mathematical modelling to solve practical problems involving additive and multiplicative situations including financial contexts; formulate the problems, choosing operations and efficient calculation strategies, using digital tools where appropriate; interpret and communicate solutions in terms of the situation	5, 7, 8, 14, 22, 24, 29, 30		
AC9M5N10	Create and use algorithms involving a sequence of steps and decisions and digital tools to experiment with factors, multiples and divisibility; identify, interpret and describe emerging patterns	9, 23, 28		
ALGEBRA				
AC9M5A01	Recognise and explain the connection between multiplication and division as inverse operations and use this to develop families of number facts	10, 15, 22, 23, 30		
AC9M5A02	Find unknown values in numerical equations involving multiplication and division using the properties of numbers and operations	6, 10, 15, 23, 32		

Australian CURRICULUM

MEASUREMENT				
AC9M5M01	Choose appropriate metric units when measuring the length, mass and capacity of objects; use smaller units or a combination of units to obtain a more accurate measure			1, 9, 13, 18, 27, 28
AC9M5M02	Solve practical problems involving the perimeter and area of regular and irregular shapes using appropriate metric units			4, 12, 16, 19, 21, 28
AC9M5M03	Compare 12- and 24-hour time systems and solve practical problems involving the conversion between them			6, 14, 23, 29
AC9M5M04	Estimate, construct and measure angles in degrees, using appropriate tools including a protractor, and relate these measures to angle names			8, 10, 22, 25, 31
SPACE				
AC9M5SP01	Connect objects to their nets and build objects from their nets using spatial and geometric reasoning			2, 11, 17, 30
AC9M5SP02	Construct a grid coordinate system that uses coordinates to locate positions within a space; use coordinates and directional language to describe position and movement			5, 7, 15, 20, 32
AC9M5SP03	Describe and perform translations, reflections and rotations of shapes, using dynamic geometric software where appropriate; recognise what changes and what remains the same, and identify any symmetries			3, 24, 26
STATISTICS				
AC9M5ST01	Acquire, validate and represent data for nominal and ordinal categorical and discrete numerical variables to address a question of interest or purpose using software including spreadsheets; discuss and report on data distributions in terms of highest frequency (mode) and shape, in the context of the data		2, 7, 10, 14, 18, 19, 23, 24, 27, 28	
AC9M5ST02	Interpret line graphs representing change over time; discuss the relationships that are represented and conclusions that can be made		11, 25, 29	
AC9M5ST03	Plan and conduct statistical investigations by posing questions or identifying a problem and collecting relevant data; choose appropriate displays and interpret the data; communicate findings within the context of the investigation		4, 8, 13, 16, 22, 32	
PROBABILITY				
AC9M5P01	List the possible outcomes of chance experiments involving equally likely outcomes and compare to those which are not equally likely		1, 3, 6, 9, 12, 30	
AC9M5P02	Conduct repeated chance experiments including those with and without equally likely outcomes, observe and record the results; use frequency to compare outcomes and estimate their likelihoods		5, 17, 20, 21, 26, 31	

Australian CURRICULUM

TARGETING MATHS HOMEWORK: YEAR 5 © PASCAL PRESS ISBN: 9781925726589

MARKING GRID

Number & Algebra: Set 1	Number & Algebra: Quick check	Number & Algebra: Set 2	Statistics & Probability	Measurement & Space	UNIT	Number & Algebra: Set 1	Number & Algebra: Quick check	Number & Algebra: Set 2	Statistics & Probability	Measurement & Space
					1					
					2					
					3					
					4					
					5					
					6					
					7					
					8					
	Units 1–2		Units 3–4		TERM 1 REVIEW		Units 5–6		Units 7–8	
					9					
					10					
					11					
					12					
					13					
					14					
					15					
					16					
	Units 9–10		Units 11–12		TERM 2 REVIEW		Units 13–14		Units 15–16	
					17					
					18					
					19					
					20					
					21					
					22					
					23					
					24					
	Units 17–18		Units 19–20		TERM 3 REVIEW		Units 21–22		Units 23–24	
					25					
					26					
					27					
					28					
					29					
					30					
					31					
					32					
	Units 25–26		Units 27–28		TERM 4 REVIEW		Units 29–30		Units 31–32	

- **Green** = Excellent — 2 or fewer questions incorrect
- **Orange** = Passing — 50% or more questions answered correctly
- **Red** = Struggling — fewer than 50% correct and needs help

Transfer your results from each unit to the grid above. Colour the traffic lights red, orange or green.

AC9M5N01

TERM 1

Decimal Numbers and Algebra

When you divide a grid into **10** parts, each part is called **one-tenth.**

This can be written as a fraction, $\frac{1}{10}$, or as a decimal, 0·1.

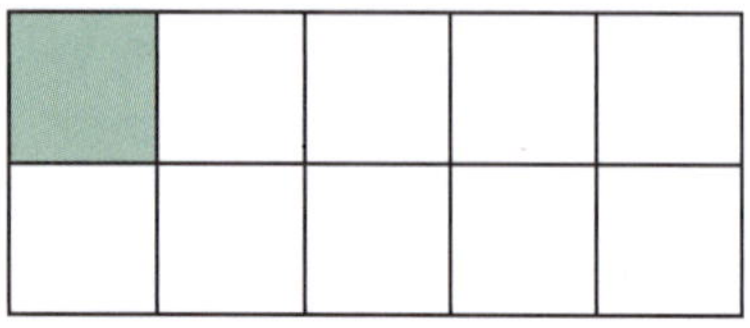

When you divide a grid into **100** parts, each part is called **one-hundredth.**

This can be written as a fraction, $\frac{1}{100}$, or as a decimal, 0·01.

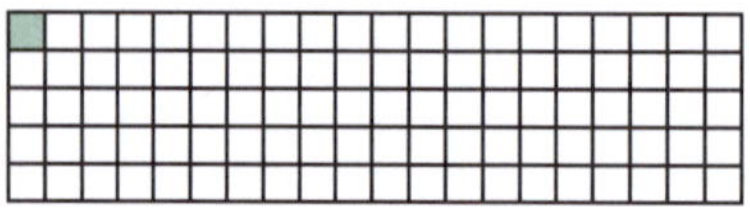

When you divide a grid into **1000** parts, each part is called **one-thousandth.**

This can be written as a fraction, $\frac{1}{1000}$, or as a decimal, 0·001.

Show these fractions on the decimal place value grid.

	Fraction	Ones	·	Tenths	Hundredths	Thousandths
1	$\frac{1}{100}$	0	·	0	1	
2	$\frac{19}{100}$	0	·	1	9	
3	$\frac{8}{1000}$					
4	$\frac{23}{100}$					
5	$\frac{13}{1000}$					
6	$\frac{143}{1000}$					
7	$\frac{12}{10}$					
8	$\frac{120}{100}$					
9	$\frac{1200}{1000}$					
10	$\frac{156}{1000}$					

SCORE /10 0-4 5-8 9-10

Quick Check

Write these decimals as fractions.

1. 0·1 = ——
2. 0·03 = ——
3. 0·004 = ——
4. 0·9 = ——
5. 0·29 = ——
6. 0·235 = ——
7. 0·165 = ——
8. 0·016 = ——
9. 0·73 = ——
10. 0·745 = ——

SCORE /10 0-4 5-8 9-10

TARGETING MATHS HOMEWORK: YEAR 5 © PASCAL PRESS ISBN: 9781925726589

Number & Algebra

AC9M5N01

Renaming Decimals for Addition

When adding decimals, it is useful to rename the decimals as fractions.
For 0·6 + 0·6, you can think:
"That is the same as $\frac{6}{10} + \frac{6}{10}$ which is $\frac{12}{10}$ or 1·2."

Complete these decimal additions.

1. 0·7 + 0·5 = $\frac{\quad}{10} + \frac{\quad}{10} = \frac{\quad}{10}$ or __·____
2. 0·9 + 0·3 = $\frac{\quad}{10} + \frac{\quad}{10} = \frac{\quad}{10}$ or __·____
3. 0·06 + 0·09 = —— + —— = —— or __·____
4. 0·07 + 0·08 = —— + —— = —— or __·____
5. 0·008 + 0·009 = —— + —— = —— or __·____
6. 0·9 + 0·7 = —— + —— = —— or __·____
7. 0·006 + 0·005 = —— + —— = —— or __·____
8. 0·07 + 0·06 = —— + —— = —— or __·____
9. 0·008 + 0·002 = —— + —— = —— or __·____
10. 0·08 + 0·03 = —— + —— = —— or __·____

SCORE /10 0-4 5-8 9-10

Statistics & Probability

AC9M5P01

Describing the Chance

We can use numbers to describe the chance of something happening.
When you are choosing between two possible outcomes, such as the three of spades or the three of hearts, we say that there is a 1 in 2 chance of the outcome being the three of hearts and a 1 in 2 chance of the outcome being the three of spades.
Each outcome has a **1 in 2** or $\frac{1}{2}$ chance of occurring.

Use numbers to describe the chance of picking a red card, without looking, from these combinations of red (r) and black (b) cards.

1. For {r, b}, there is a 1 in 2 or $\frac{1}{2}$ chance of picking red.
2. For {b, r, r}, there is a __ in __ or —— chance of picking red.
3. For {b, r, r, b}, there is a __ in __ or —— chance of picking red.
4. For {r, b, b, b}, there is a __ in __ or —— chance of picking red.
5. For {r, r, b, b, r}, there is a __ in __ or —— chance of picking red.

SCORE /5 0-2 3-4 5

Measurement & Space

AC9M5M01

Units of Measure

Which units would you use to measure these items?
Write the full name of the unit and the abbreviation, e.g. centimetres and cm.

	Item to measure	Name of the unit	Abbreviated form
1	The width of the oval		
2	The mass of an apple		
3	The area of this page		
4	The capacity of a teacup		
5	The mass of a box of apples		
6	The length of a pencil		
7	The capacity of a bathtub		
8	Something shorter than a centimetre		
9	The distance between two towns		
10	Your height		
11	Your mass		
12	Your age		

SCORE /12 0-5 6-10 11-12

NAPLAN*-style Questions

1. What is the next number in this sequence?

   0·002, 0·004, 0·006, 0·008, ________

   ◯ 0·0001 ◯ 0·01 ◯ 0·0010 ◯ 0·001

2. Which of these spinners is most likely to land on green?

   ◯ ◯ ◯ ◯

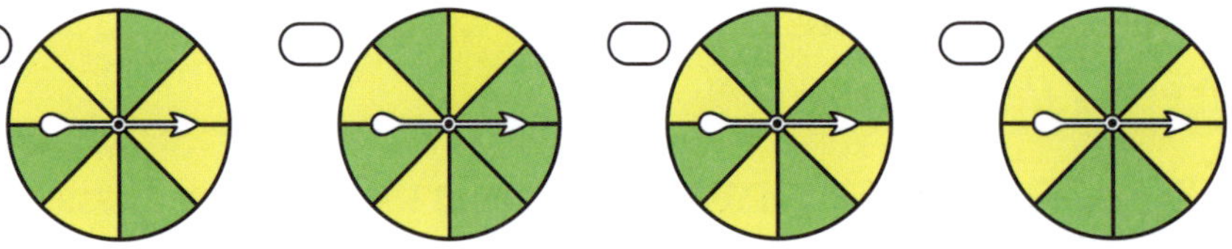

3. Miro needs 1·25 kg of sugar to make marmalade. The scales show that he doesn't have enough.

   How much more sugar does Miro need to make the marmalade?

   ________ kg

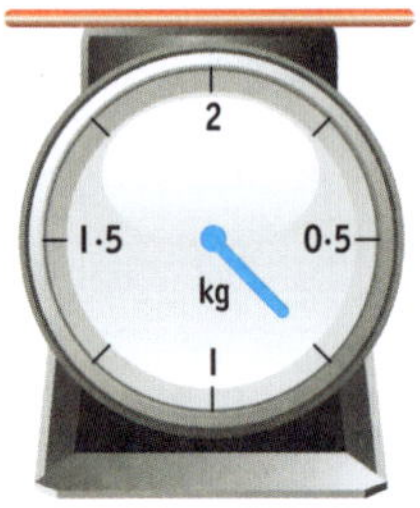

SCORE /3 0-1 2 3

TARGETING MATHS HOMEWORK: YEAR 5 © PASCAL PRESS ISBN: 9781925726589

Problem of the Week

AC9M5P01

Just by Chance

We took these four cards from a pack of cards:

We placed them facedown on the table and shuffled them around.

Then we picked one of the cards to turn over.

We repeated this experiment 12 times.

① What do you think will happen if you repeat their experiment 12 times?

____

Try the experiment and see what the outcome is.
If you don't have a pack of cards, use four pieces of paper labelled like this:

Red Heart	Red Diamond	Black Club	Black Spade

Record your results in the table.

	Red		Black	
	Heart	Diamond	Club	Spade
1				
2				
3				
4				
5				
6				
7				
8				
9				
10				
11				
12				

Record the outcomes.

② Red was chosen ____ in 12 times.

③ Black was chosen ____ in 12 times.

④ Heart was chosen ____ in 12 times.

⑤ Diamond was chosen ____ in 12 times.

⑥ Club was chosen ____ in 12 times.

⑦ Spade was chosen ____ in 12 times.

AC9M5N02

Factors

6 can be made as 1 × 6, 6 × 1, 2 × 3 and 3 × 2.
The numbers 1, 2, 3 and 6 are called the **factors** of 6.
All whole numbers have themselves and 1 as factors.

5 can only be made as 1 × 5 and 5 × 1.
So 5 has only two factors, 1 and 5.

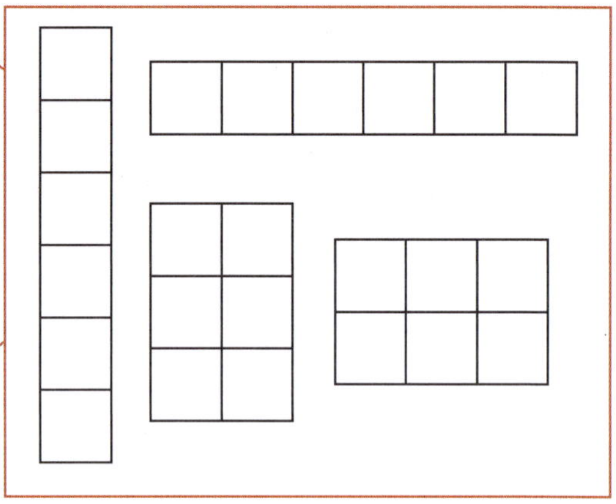

Show the factors of 12 by drawing rectangles with an area of 12 squares on this grid. Write the factors of 12 that each rectangle makes. Score 1 point for each rectangle.

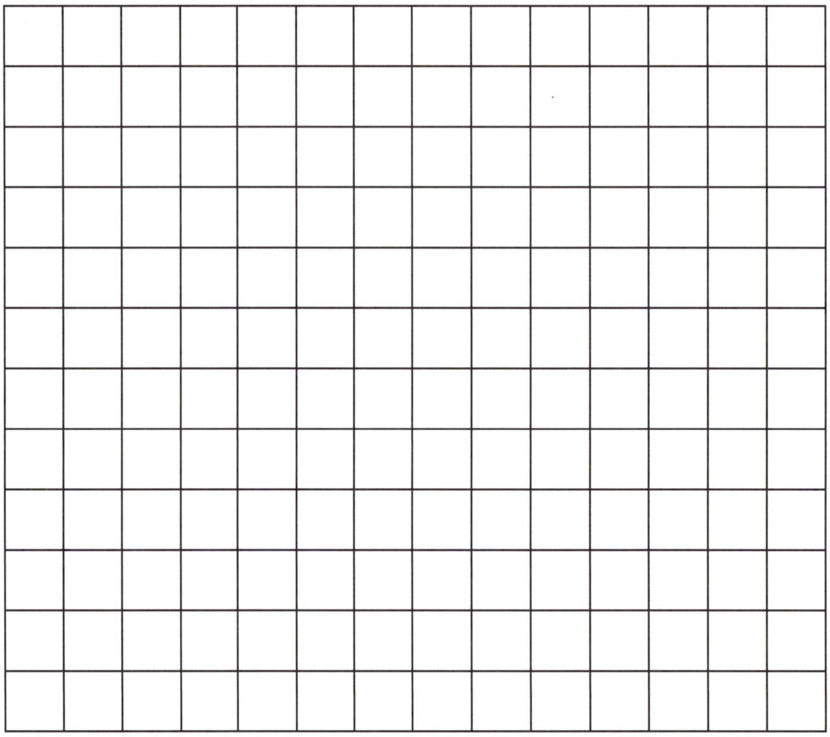

SCORE /6 0-2 3-4 5-6

Quick Check

Write the factors.

1. The factors of 2 are __ and __.
2. The factors of 3 are __ and __.
3. The factors of 4 are __, __ and __.
4. The factors of 7 are __ and __.
5. The factors of 8 are __, __, __ and __.
6. The factors of 9 are __, __ and __.
7. The factors of 10 are __, __, __ and ___.
8. The factors of 11 are __ and ___.
9. The factors of 14 are __, __, __ and ___.
10. The factors of 15 are __, __, __ and ___.

SCORE /10 0-4 5-8 9-10

TARGETING MATHS HOMEWORK: YEAR 5 © PASCAL PRESS ISBN: 9781925726589

Number & Algebra

AC9M5N02

Multiples

When we multiply two whole numbers, the answer is a multiple of the two numbers.

$2 \times 3 = 6$

2 and 3 are **factors** of 6.

6 is a **multiple** of 2 and a **multiple** of 3.

For each number in the first column, tick the numbers in the other columns that are factors. The first one has been done for you. It shows that 2, 3, 4 and 6 are factors of 12.

	Number	2	3	4	5	6	7	8	9
1	12	✓	✓	✓		✓			
2	15								
3	20								
4	24								
5	28								

	Number	2	3	4	5	6	7	8	9
6	30								
7	36								
8	42								
9	45								
10	56								

SCORE /10 0-4 5-8 9-10

Statistics & Probability

AC9M5ST01

Show Ratings

Leo asked six people to rate their favourite types of TV shows out of 5, where 1 was the least favourite and 5 was the most favourite. He called his six voters **A**, **B**, **C**, **D**, **E** and **F**.

They could vote for **Sit-com** (situation comedy shows), **Doco** (documentary TV shows) and **Sci-fi** (science fiction shows). He showed his data in two ways:

	Voters					
TV Show	A	B	C	D	E	F
Sit-com	3	3	4	2	5	3
Doco	4	4	2	5	5	1
Sci-fi	2	2	3	4	3	2

	Ratings				
TV Show	1	2	3	4	5
Sit-com		✓	✓✓✓	✓	✓
Doco	✓	✓		✓✓	✓✓
Sci-fi		✓✓✓	✓✓	✓	

Answer these questions and say which table (Voters or Ratings) you used to answer that question.

		Answer	Table
1	What was the total rating for documentaries?		
2	Which rating score was the least common?		
3	Which TV show category scored least?		
4	What was the range of scores for Sci-fi?		
5	Which TV show category had the largest range of scores?		

SCORE /5 0-2 3-4 5

Measurement & Space

AC9M5SP01

Made with Shapes

Jade made this tower with two 3D objects that she had made from nets. What are the names of the two 3D objects that she made?

1. ______________________
2. ______________________

Circle the nets Jade could have used.

3. 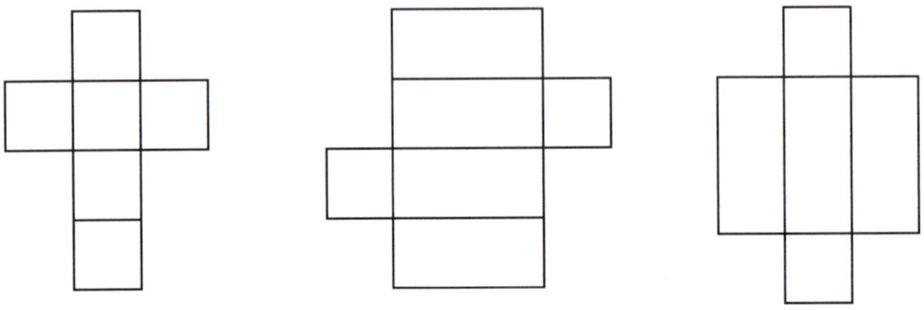

4. 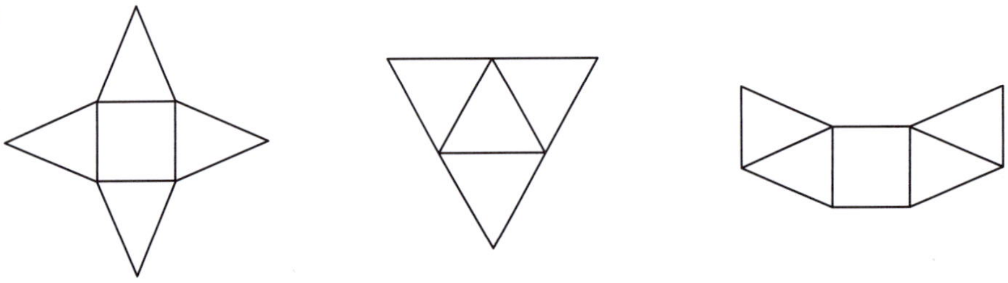

If Jade wants to make a round tower, which two 3D objects should she use?

5. Jade should use a ____________ and a ______________.

SCORE | /5 | 0-2 | 3-4 | 5

NAPLAN*-style Questions

1. Which of these numbers is a multiple of 2, 3 and 4?
   - ◯ 4 ◯ 6 ◯ 8 ◯ 12

2. This is the net of a 3D object.

   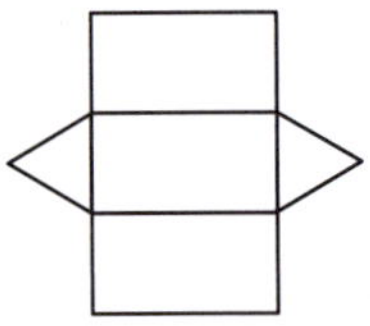

   Which of these 3D objects would it make?

   ◯ 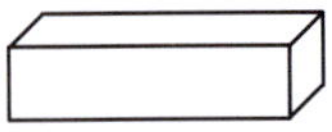 ◯  ◯ 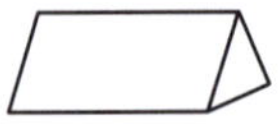 ◯ 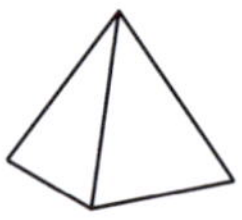

3. Which of these numbers is not a factor of 21?
   - ◯ 1 ◯ 3 ◯ 7 ◯ 9

SCORE | /3 | 0-1 | 2 | 3

TARGETING MATHS HOMEWORK: YEAR 5 © PASCAL PRESS ISBN: 9781925726589

Shape Puzzle

"Class, I have prepared a set of 6 nets for each group to fold and glue to make 3D objects.

I look forward to seeing your results."

Mrs Green was very pleased to see the 3D objects that Asha's group had made and labelled so neatly.

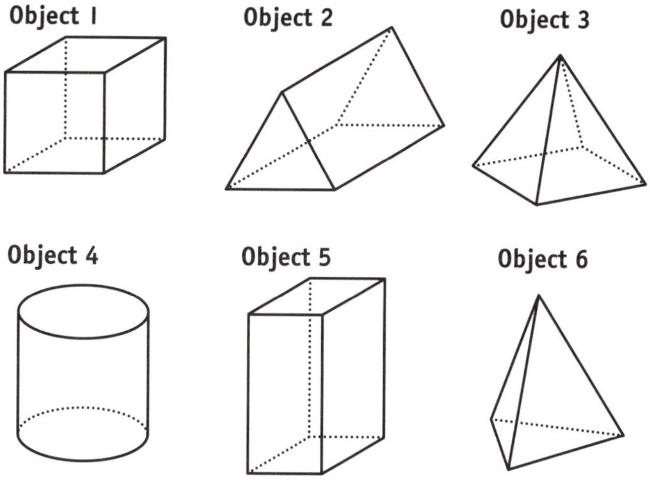

"Mrs Green," said Asha, "Sam, Leo, Jade, Miro and Ravi are in my group, and we have all made up a clue so that the other groups can guess which 3D object we each made."

"What a wonderful idea, Asha. I will put your clues to the rest of the class straight away," said Mrs Green.

Miro's Clue:
My net had five shapes and two of them were the same and the other three were also the same.

Sam's Clue:
My net had five shapes and four of them were identical.

Jade's Clue:
My net did not have any circles in it.

Asha's Clue:
My net made the shape that no-one else made.

Ravi's Clue:
My net had six shapes and they were all the same.

Leo's Clue:
My net had shapes that were all the same and none of them had sides that were parallel.

Who made which object?

Object 1: ______________

Object 2: ______________

Object 3: ______________

Object 4: ______________

Object 5: ______________

Object 6: ______________

AC9M5N03

Ordering Mixed Fractions

When you want to put fractions like $1\frac{1}{8}$, $\frac{5}{8}$ and $\frac{13}{8}$ in order of size, you need to convert the mixed number into eighths: $\frac{9}{8}$, $\frac{5}{8}$ and $\frac{13}{8}$.
Now you can place the fractions on a number line and the order of the fractions becomes clear:

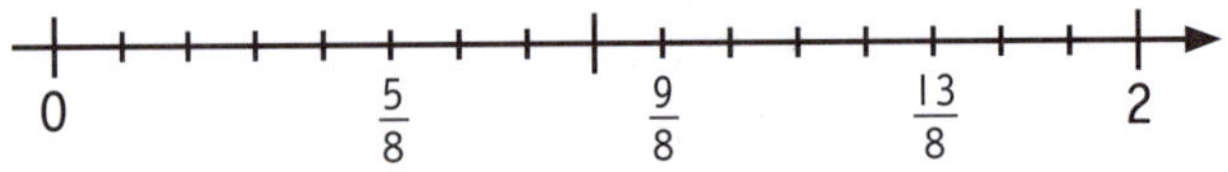

Convert the mixed numbers and show the fractions on the number lines.

1. $1\frac{3}{5}$, $\frac{6}{5}$, $\frac{9}{5}$

   ___, ___, ___

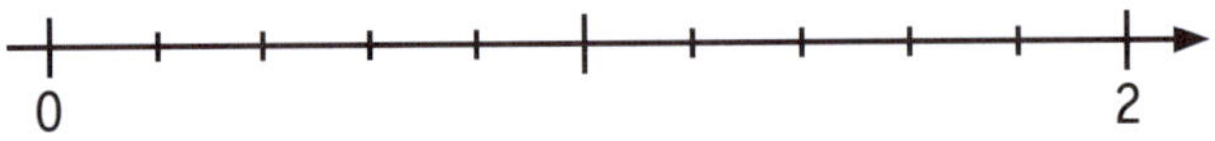

2. $\frac{1}{2}$, $1\frac{1}{4}$, $\frac{7}{4}$

   ___, ___, ___

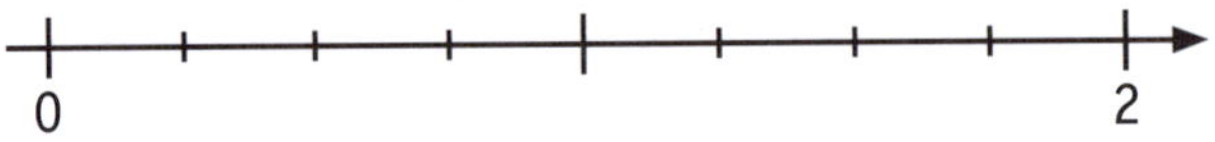

3. $\frac{1}{6}$, $1\frac{1}{6}$, $\frac{11}{6}$

   ___, ___, ___

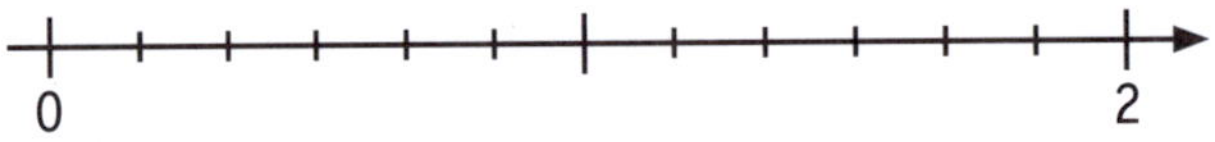

4. $\frac{12}{4}$, $1\frac{1}{4}$, $2\frac{3}{4}$

   ___, ___, ___

5. $\frac{5}{3}$, $1\frac{1}{3}$, $\frac{9}{3}$

   ___, ___, ___

SCORE /5 0-2 3-4 5

Quick Check

Circle the larger fraction in each pair.

1. $1\frac{1}{8}$ $\frac{11}{8}$
2. $1\frac{1}{5}$ $\frac{7}{5}$
3. $\frac{15}{8}$ $1\frac{6}{8}$
4. $\frac{14}{10}$ $1\frac{6}{10}$
5. $\frac{17}{6}$ $3\frac{1}{6}$
6. $3\frac{1}{4}$ $\frac{15}{4}$
7. $1\frac{3}{5}$ $\frac{9}{5}$
8. $\frac{13}{2}$ $5\frac{1}{2}$
9. $\frac{16}{3}$ $4\frac{2}{3}$
10. $1\frac{3}{7}$ $\frac{15}{7}$

SCORE /10 0-4 5-8 9-10

TARGETING MATHS HOMEWORK: YEAR 5 © PASCAL PRESS ISBN: 9781925726589

Number & Algebra

AC9M5N05

Fractions and Addition

When we add two fractions such as $\frac{3}{4} + \frac{1}{8}$, we can see that the two fraction pieces for quarters and eighths are not the same size.

But what if we redraw the $\frac{3}{4}$ as $\frac{6}{8}$?

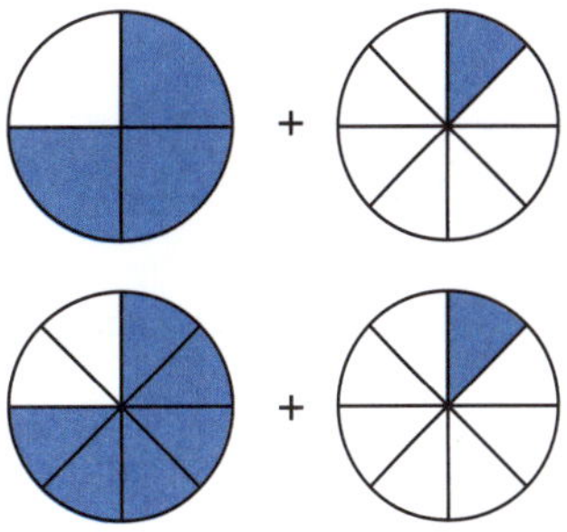

Now the pieces are all the same size, and we see that there are seven $\frac{1}{8}$ pieces altogether.

We can also show this with numbers: $\frac{3}{4} + \frac{1}{8} = \frac{6}{8} + \frac{1}{8} = \frac{7}{8}$

Convert these fractions before adding them.

1. $\frac{1}{6} + \frac{2}{3} = \frac{1}{6} + \frac{}{6} = \frac{}{6}$
2. $\frac{1}{2} + \frac{3}{4} = \frac{}{4} + \frac{}{} = \frac{}{}$
3. $\frac{5}{6} + \frac{1}{3} = \frac{}{} + \frac{}{} = \frac{}{}$
4. $\frac{5}{8} + \frac{1}{4} = \frac{}{} + \frac{}{} = \frac{}{}$
5. $\frac{1}{6} + \frac{1}{2} = \frac{}{} + \frac{}{} = \frac{}{}$
6. $\frac{3}{8} + \frac{1}{4} = \frac{}{} + \frac{}{} = \frac{}{}$
7. $\frac{3}{5} + \frac{2}{10} = \frac{}{} + \frac{}{} = \frac{}{}$
8. $\frac{1}{2} + \frac{3}{8} = \frac{}{} + \frac{}{} = \frac{}{}$
9. $\frac{2}{6} + \frac{1}{2} = \frac{}{} + \frac{}{} = \frac{}{}$
10. $\frac{3}{10} + \frac{1}{2} = \frac{}{} + \frac{}{} = \frac{}{}$
11. $\frac{1}{3} + \frac{5}{6} = \frac{}{} + \frac{}{} = \frac{}{}$
12. $\frac{1}{6} + \frac{5}{3} = \frac{}{} + \frac{}{} = \frac{}{}$

SCORE /12 0-5 6-10 11-12

Statistics & Probability

AC9M5P01

Spin a Number

Jade and Leo have each made a spinner.

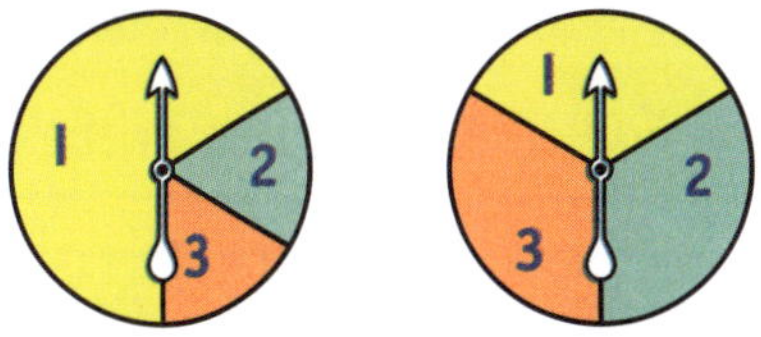

Jade's spinner   Leo's spinner

1. Whose spinner has the best chance of spinning a 1?

   ______

2. Whose spinner has an equal chance of spinning a 1, a 2 or a 3?

   ______

3. If Jade's spinner was spun 6 times, how many times do you think a 2 would be spun? ____
4. Why do you think that? ______________________________
5. If Leo's spinner was spun 6 times, how many times do you think a 2 would be spun? ____
6. Why do you think that? ______________________________

SCORE /6 0-2 3-4 5-6

AC9M5SP03

Slides, Flips and Turns

I have been using bread tags to practise transformations.
I painted one side orange and the other side blue and used the tags to show the types of transformations that you can make.

Slide 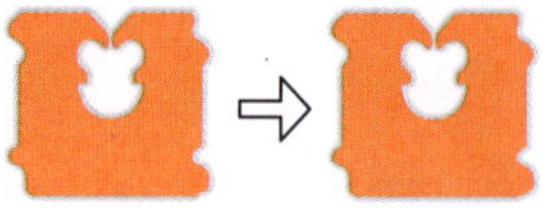 I made this by sliding the bread tag in one direction.

Flip 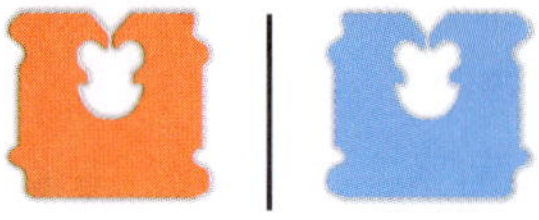 I made this by flipping the bread tag over.

Turn  I made this by turning the bread tag clockwise through an angle of 90°.

Draw and colour the next bread tag in these patterns.

① 

The transformation is a ________.

② 

The transformation is a ________.

③ 

The transformation is a ________.

④ 

The transformation is a ________.

⑤ 

The transformations are a ________ and a ________.

SCORE /5 0-2 3-4 5

NAPLAN*-style Questions

① Circle the fraction in the box that the ? represents.

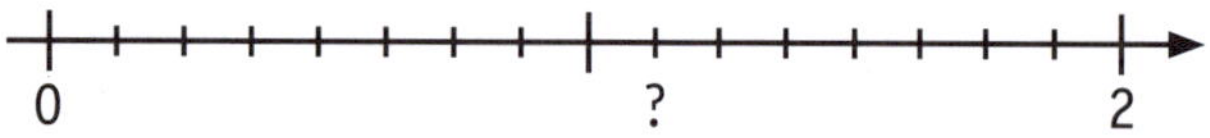

$\frac{9}{16}$	$1\frac{1}{16}$	$\frac{9}{9}$	$1\frac{1}{8}$

② Circle the bread tag in the box that will come next in this pattern.

③ Which option does not represent the chance of spinning purple with this spinner?

- ⬭ 3 in 6
- ⬭ even
- ⬭ $\frac{1}{3}$
- ⬭ quite likely

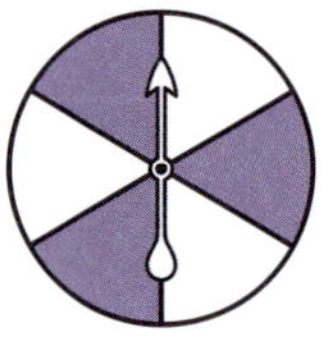

SCORE /3 0-1 2 3

TARGETING MATHS HOMEWORK: YEAR 5 © PASCAL PRESS ISBN: 9781925726589

Move It

I have been working with Ravi to make up a problem using the bread tags I made.

You can move a bread tag on a grid in two ways.

1 Either **slide** it one space left or right or one space up or down.

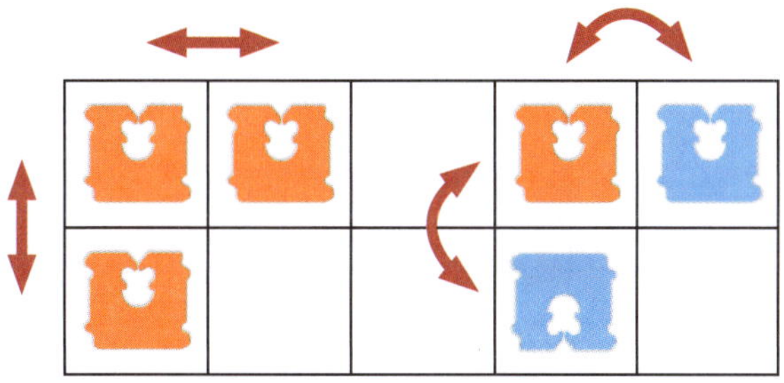

2 Or **flip** the bread tag over a side of the square that you are in.

Here is our problem:

Move the bread tag from the Start position to the Finish position in just 10 moves.

You have to use as many slide moves as you make flip moves.

Start

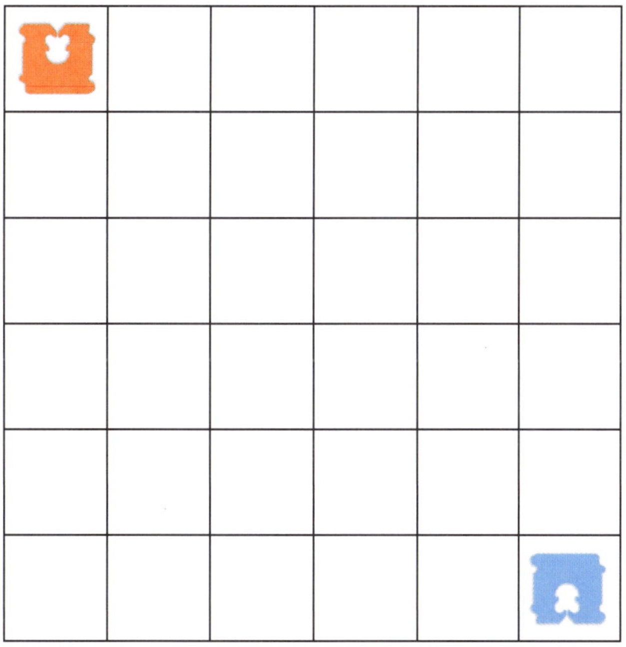

Finish

A Further Investigation

Suppose you make an **even number** of moves from the Start position.

Which squares of the grid can you reach?

If you make an **odd number** of moves, which squares can you reach?

Number & Algebra

AC9M5N04

Percentages

Per cent means out of 100. A percentage is a fraction of 100 and a form of decimal number.

These grids show how percentages, fractions and decimals are related.

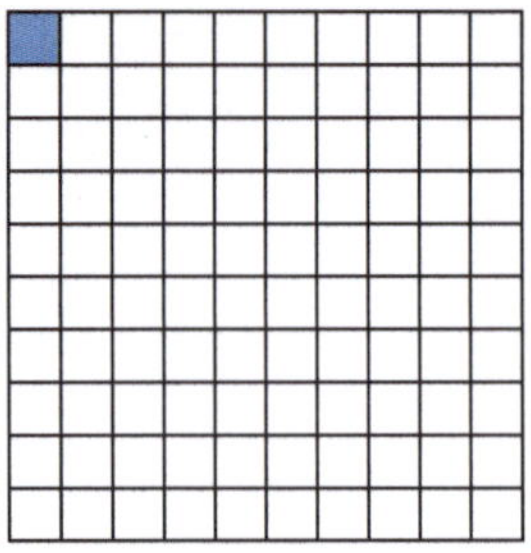

1% Percentage, $\frac{1}{100}$ Fraction, 0·01 Decimal

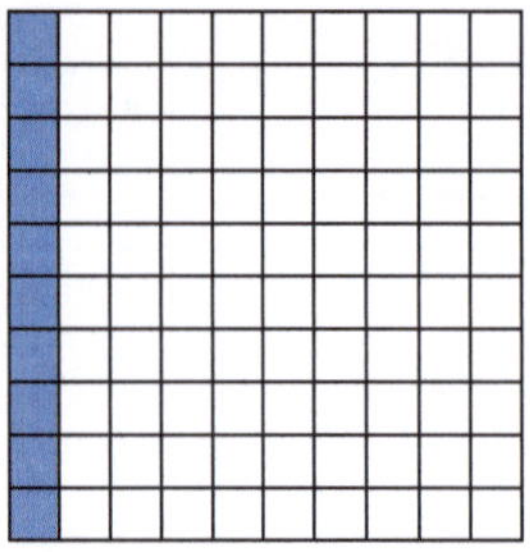

10% Percentage, $\frac{10}{100}$ Fraction, 0·1 Decimal

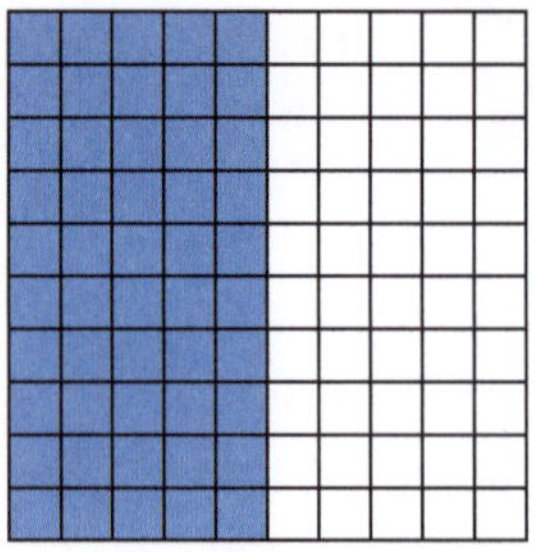

50% Percentage, $\frac{50}{100}$ Fraction, 0·50 Decimal

Write the percentages, fractions and decimals for the blue squares in these grids.

1. 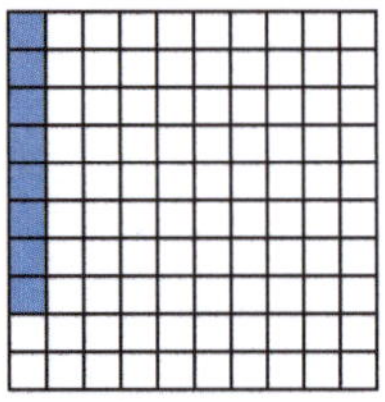
   ____% Percentage
   —— Fraction
   0·____ Decimal

2. 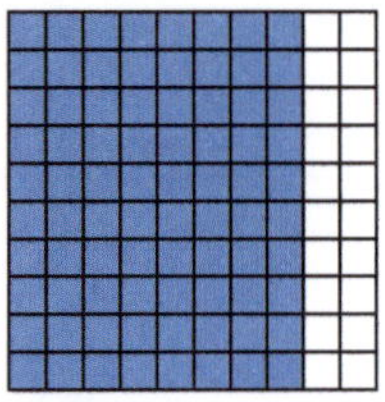
   ____% Percentage
   —— Fraction
   0·____ Decimal

3. 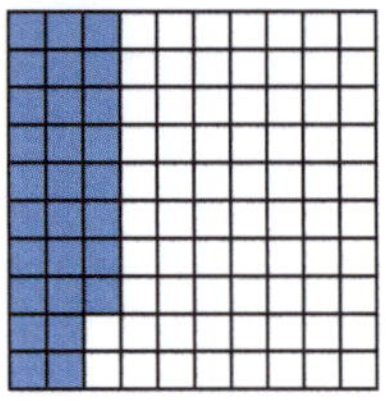
   ____% Percentage
   —— Fraction
   0·____ Decimal

4. 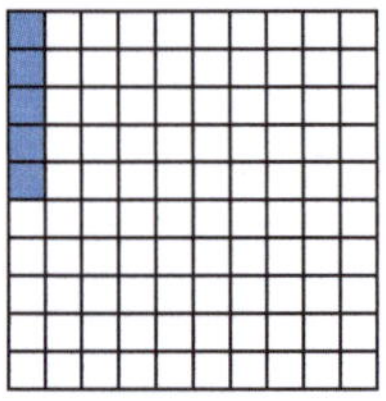
   ____% Percentage
   —— Fraction
   0·____ Decimal

5. 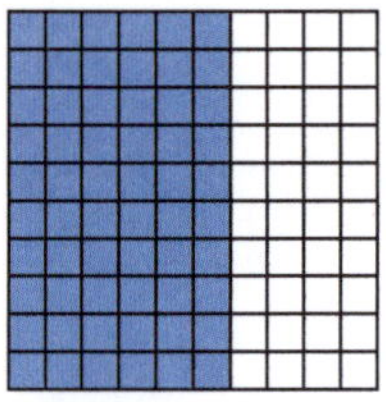
   ____% Percentage
   —— Fraction
   0·____ Decimal

6. 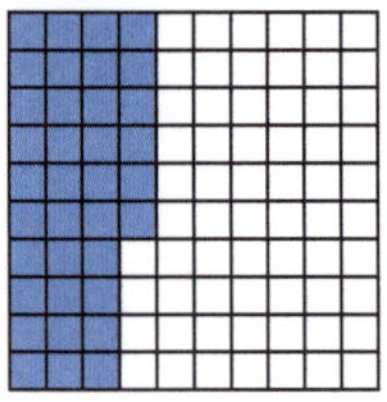
   ____% Percentage
   —— Fraction
   0·____ Decimal

SCORE /6 0-2 3-4 5-6

Quick Check

Write these fractions as decimals and percentages.

1. $\frac{3}{100}$ is the same as 0·____ and ____%.
2. $\frac{22}{100}$ is the same as 0·____ and ____%.
3. $\frac{5}{100}$ is the same as 0·____ and ____%.
4. $\frac{17}{100}$ is the same as 0·____ and ____%.

Write these percentages as decimals and fractions.

5. 18% is the same as 0·____ and ——.
6. 6% is the same as 0·____ and ——.
7. 45% is the same as 0·____ and ——.
8. 9% is the same as 0·____ and ——.

SCORE /8 0-3 4-6 7-8

TARGETING MATHS HOMEWORK: YEAR 5 © PASCAL PRESS ISBN: 9781925726589

Number & Algebra

Percentage Discounts

I like sale time because the things I need can be bought at special prices.

During a sale, a $100 skateboard might have a 5% discount.

5% = 0·05 or $\frac{5}{100}$

5 hundredths of $100 is $5, so the skateboard will now cost:
$100 – $5 = $95.

What would a $100 skateboard cost with these discounts?

1. 20% = 0·____ or $\frac{\quad}{100}$ The discount is $____ and the skateboard would cost $____.
2. 15% = 0·____ or $\frac{\quad}{100}$ The discount is $____ and the skateboard would cost $____.
3. 7% = 0·____ or $\frac{\quad}{100}$ The discount is $____ and the skateboard would cost $____.
4. 50% = 0·____ or $\frac{\quad}{100}$ The discount is $____ and the skateboard would cost $____.
5. 30% = 0·____ or $\frac{\quad}{100}$ The discount is $____ and the skateboard would cost $____.
6. 10% = 0·____ or $\frac{\quad}{100}$ The discount is $____ and the skateboard would cost $____.
7. 12% = 0·____ or $\frac{\quad}{100}$ The discount is $____ and the skateboard would cost $____.
8. 9% = 0·____ or $\frac{\quad}{100}$ The discount is $____ and the skateboard would cost $____.

SCORE /8 0-3 4-6 7-8

Statistics & Probability

The Fruit Survey

100 people were surveyed about their preferred fruit.

- 25% chose bananas.
- 10% chose apples.
- 30% chose strawberries.
- The rest chose mandarins.

1. Colour the graph to show the results.
   Use:
   - yellow for banana
   - green for apple
   - red for strawberry
   - orange for mandarin.
2. Which fruit was the most popular? __________
3. How many people chose bananas? _____
4. What percentage of the people chose mandarins? _____
5. What percentage of the people did not choose mandarins? _____

SCORE /5 0-2 3-4 5

Measurement & Space

AC9M5M02

Area and Perimeter

To find the area of this rectangle you could count every square:

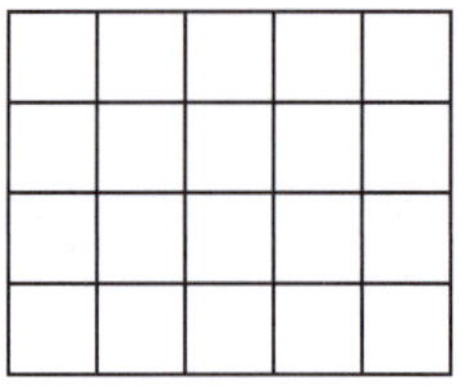

Each square represents 1 m by 1 m.

But that would be TOO SLOW! It would be much better to count how many squares across and how many squares down.

There are 5 squares across and 4 squares down.

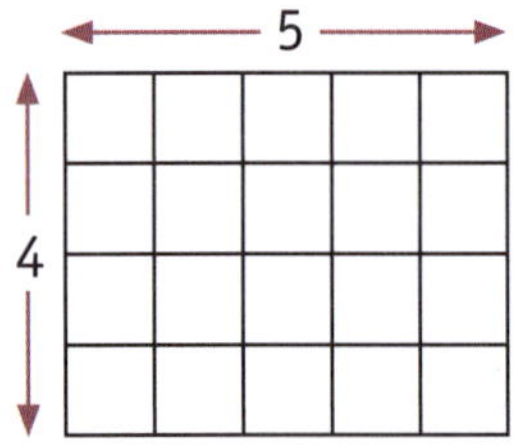

Multiply 5 m × 4 m and you can see that the area is 20 m^2.

Now that you have found the area, the same information can be used to find the perimeter.

5 m across, 4 m down, 5 m back across and 4 m up to the start position.

This makes the perimeter: 5 + 4 + 5 + 4 = 18 m.

Use multiplication to find the area and addition to find the perimeter of these shapes, which are made of 1 m × 1 m squares.

1. 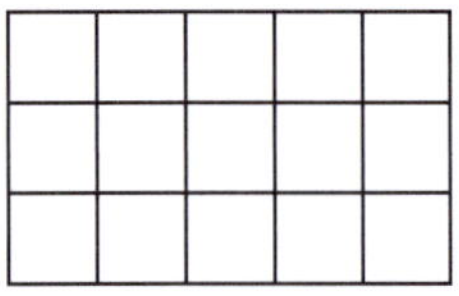

   Area = _____ m^2 Perimeter = _____ m

2. 

   Area = _____ m^2 Perimeter = _____ m

3. 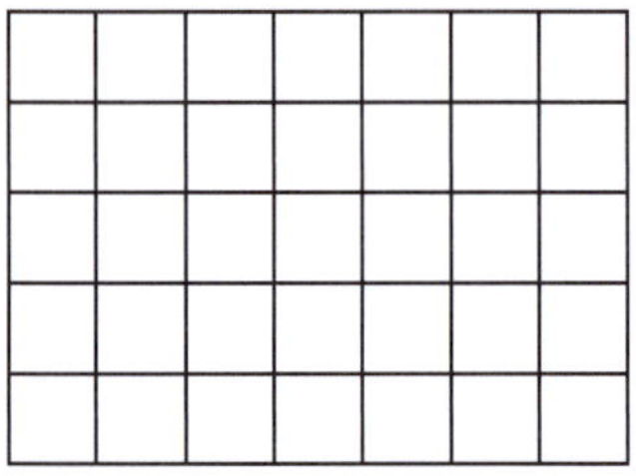

   Area = _____ m^2 Perimeter = _____ m

4. 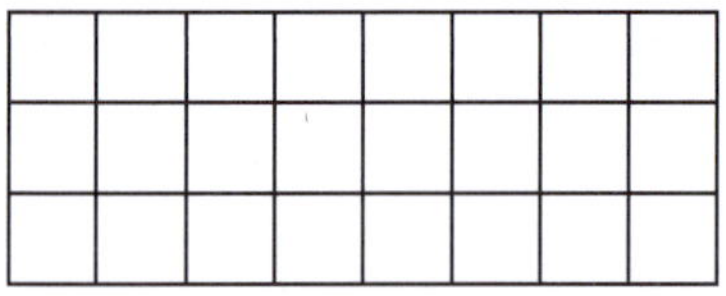

   Area = _____ m^2 Perimeter = _____ m

SCORE /4 (0-1) (2-3) (4)

NAPLAN*-style Questions

1. Which of these decimals is equivalent to 23%?

   ◯ 2·3 ◯ 0·23 ◯ 0·023 ◯ 23

2. What is the area and the perimeter of this rectangle?

   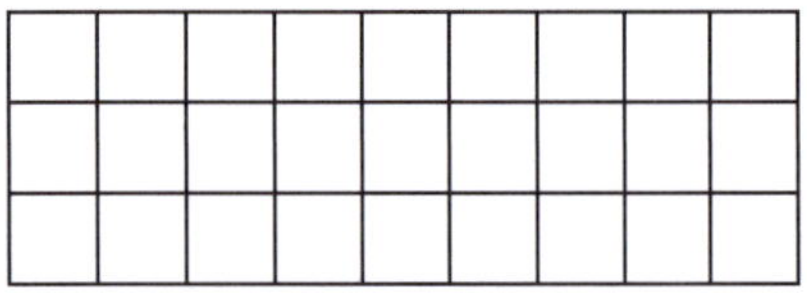

   Area = _____ m^2

   Perimeter = _____ m

3. What percentage of the pie chart is yellow?

   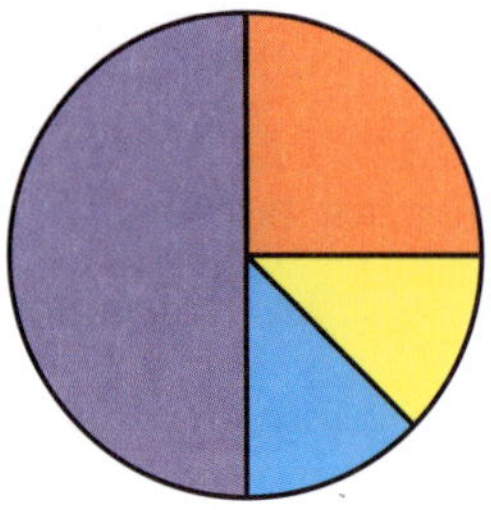

   [ ] %

SCORE /3 (0-1) (2) (3)

TARGETING MATHS HOMEWORK: YEAR 5 © PASCAL PRESS ISBN: 9781925726589

Problem of the Week

Counters on a Grid

Miro has covered a 100-square grid with counters.

- 24% of his counters are red.
- 66% of his counters are purple.
- There are the same number of yellow counters as green counters.
- Use the grid to show how many counters of each colour Miro has placed on his grid.

Ella has covered her 100-square grid too.

- 15% of her counters are purple.
- She has used 5 more red than purple counters.
- 45% of her counters are yellow. All the rest are green.
- Use this grid to show how many counters of each colour Ella has placed on her grid.

Use this grid to make up your own counters on a grid problem.

What are your clues that will let other people find out how many of each colour you placed on the grid?

Clue 1 ____

____

Clue 2 ____

____

Clue 3 ____

____

Clue 4 ____

____

Number & Algebra

AC9M5N06

Multiplication of Large Numbers

One way of multiplying large numbers by a single-digit number is to use the area model.

For 68 × 7, we can split the 68 into 60 and 8 and then multiply the parts:

	60 +	8	
			= 68
× 7	420	56	= 476

This works for larger numbers too. 368 × 7 looks like this:

	300 +	60 +	8	
				= 368
× 7	2100	420	56	= 2576

You should always check to see if your answer is reasonable.
For example, 68 × 7 should be close to but more than 60 × 7 = 420.
Also, it should be less than 70 × 7 = 490. So the answer should be between 420 and 490.

Complete these area model multiplications. Then give the estimated between values.

1. 73 × 8

   ____ + ____ = ____

   × __ [____ | ____] = ____

   Between ______ and ______.

2. 222 × 6

   ____ + ____ + ____ = ____

   × __ [____ | ____ | ____] = ____

   Between ______ and ______.

3. 36 × 5

   ____ + ____ = ____

   × __ [____ | ____] = ____

   Between ______ and ______.

4. 135 × 8

   ____ + ____ + ____ = ____

   × __ [____ | ____ | ____] = ____

   Between ______ and ______.

5. 83 × 6

   ____ + ____ = ____

   × __ [____ | ____] = ____

   Between ______ and ______.

6. 476 × 5

   ____ + ____ + ____ = ____

   × __ [____ | ____ | ____] = ____

   Between ______ and ______.

SCORE /6 0-2 3-4 5-6

Quick Check

Write the estimated values that these multiplications lie between.

1. For 45 × 6, between _____ and _____.
2. For 83 × 8, between _____ and _____.
3. For 59 × 7, between _____ and _____.
4. For 37 × 5, between _____ and _____.
5. For 64 × 6, between _____ and _____.
6. For 27 × 9, between _____ and _____.
7. For 84 × 6, between _____ and _____.
8. For 75 × 4, between _____ and _____.
9. For 93 × 3, between _____ and _____.
10. For 38 × 4, between _____ and _____.

SCORE /10 0-4 5-8 9-10

TARGETING MATHS HOMEWORK: YEAR 5 © PASCAL PRESS ISBN: 9781925726589

Number & Algebra

AC9M5N09

Multiplying Money Amounts

Instead of number splitting, you can do **amount splitting** when multiplying money amounts.

For 6 × $3·45, split the money amount into $3, 40c and 5c:

	$3 +	40c +	5c	= $3·45
× 6	$18	$2·40	30c	= $20·70

You should always check to see if your answer is reasonable.
For example, 6 × $3·45 should be more than 6 × $3 = $18.
Also, it should be less than 6 × $4 = $24.
So the answer should be between $18 and $24.

Complete these money multiplications. Then give the estimated between values.

1. 5 × $5·75

	$5 +	70c +	5c	= $5·75
× 5				= $______

Between ______ and ______

2. 7 × $2·25

	$___ +	___c +	___c	= $______
× __				= $______

Between ______ and ______

3. 4 × $6·39

	$___ +	___c +	___c	= $______
× __				= $______

Between ______ and ______

4. 6 × $8·37

	$___ +	___c +	___c	= $______
× __				= $______

Between ______ and ______

5. 3 × $12·65

	$___ +	___c +	___c	= $______
× __				= $______

Between ______ and ______

6. 9 × $3·25

	$___ +	___c +	___c	= $______
× __				= $______

Between ______ and ______

SCORE /6 0-2 3-4 5-6

Statistics & Probability

AC9M5P02

Just by Chance

Asha tossed her coin 10 times and these were her results:

Heads	|||
Tails	卌 ||

Ravi tossed his coin 10 times and these were his results:

Heads	卌 |
Tails	||||

1. How many times out of 10 did Asha toss a tail? ___
2. What were Ravi's results? Heads: ___ out of ___. Tails: ___ out of ___.
3. If they repeated the process, would they get the same results next time? _______
4. Why or why not? ______________________________
5. Is the statement "Tails has a 5 in 10 chance" reasonable? ______________________

   Why or why not? ______________________

SCORE /5 0-2 3-4 5

AC9M5SP02

Coordinate Grids

On a coordinate grid like this one, the numbers are on the lines, not in the spaces.

The coordinates of a point are a pair of numbers inside brackets and separated by a comma.

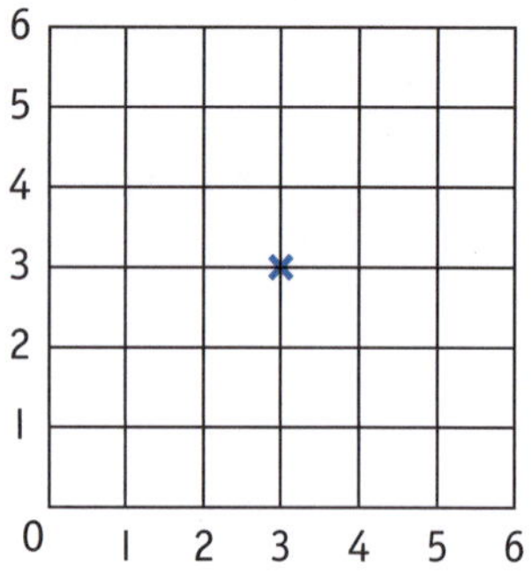

The blue cross marks (3, 3) on the grid.

Across (Horizontal) Up (Vertical)

① Draw a black cross at each intersection around the blue cross to make a square.

② Record the coordinates of each of the crosses that you made.
Don't forget the brackets and the commas.

_______ _______ _______ _______ _______ _______ _______ _______

③ Use a different colour to mark the intersections around the square to make a new, larger square.

④ Record the coordinates of the points that make up the new square.

_______ _______ _______ _______ _______ _______ _______ _______

_______ _______ _______ _______ _______ _______ _______ _______

⑤ Draw a line from (1, 5) to (5, 1) and a line from (1, 1) to (5, 5) to make a cross.

SCORE /5 0-2 3-4 5

NAPLAN*-style Questions

① Sam bought 6 bananas at $1·35 each.

How much did she pay for the bananas? __________

② What are the chances of spinning orange with this spinner?

◯ 3 in 6 ◯ 3 in 8 ◯ 5 in 8 ◯ 5 in 6

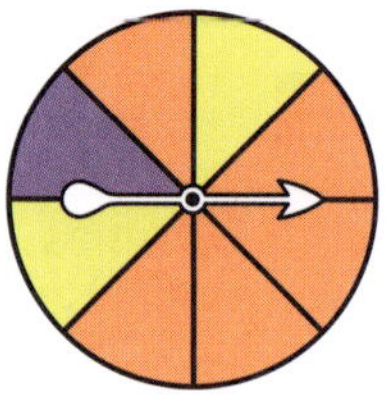

③ What are the coordinates of the point marked ■?

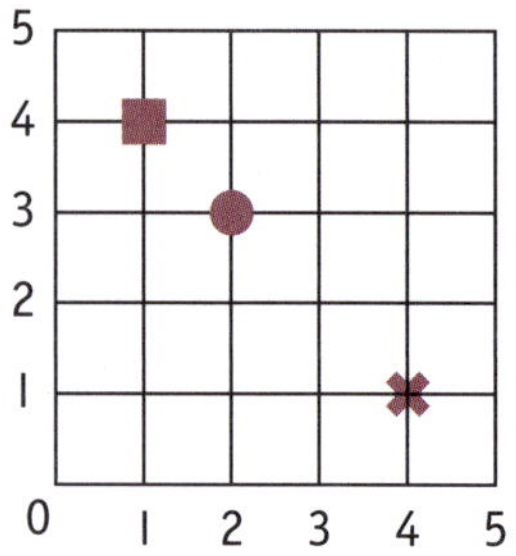

◯ (4, 1)
◯ (0, 4)
◯ (1, 4)
◯ (4, 4)

SCORE /3 0-1 2 3

TARGETING MATHS HOMEWORK: YEAR 5 © PASCAL PRESS ISBN: 9781925726589

Problem of the Week

AC9M5N06

Prizes for a Lucky Dip

We all went shopping yesterday to get prizes for the lucky dip.

Jade: "I spent the least."

Leo: "I spent the most."

Miro: "I spent $1·05 more than Jade."

Ella: "I'm not telling!"

This is what we bought:

3 pencils at $7·45 each

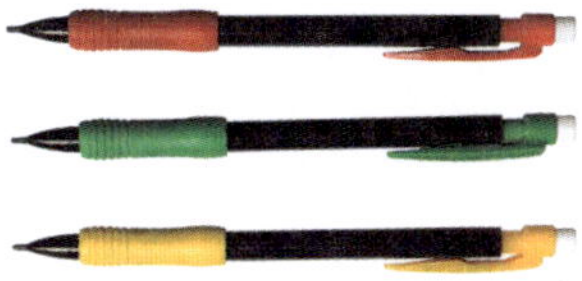

6 bouncy balls at $3·55 each

4 notebooks at $7·45 each

5 pencil sharpeners at $5·39 each

Use these multiplication grids to help you work out who bought which Lucky Dip prizes.

	$___ +	___c +	___c	= $________
× ___				= $________

	$___ +	___c +	___c	= $________
× ___				= $________

	$___ +	___c +	___c	= $________
× ___				= $________

	$___ +	___c +	___c	= $________
× ___				= $________

Who spent what?

Jade spent $__________ on ________________________________.

Leo spent $__________ on ________________________________.

Miro spent $__________ on ________________________________.

Ella spent $__________ on ________________________________.

Division and Related Multiplication Facts

If I was asked, "What is 54 ÷ 9?", I would think about related multiplication facts: "What multiplied by 9 makes 54?"
The answer is 6 × 9 = 54.
So, when I divide 54 by 9, I know that the answer is 6.

Write the related multiplication facts and answers for these division questions.

1. 36 ÷ 9 = _____
   _____ × 9 = 36
2. 63 ÷ 7 = _____
   _____ × 7 = 63
3. 56 ÷ 7 = _____
   _____ × 7 = 56
4. 42 ÷ 6 = _____
   _____ × 6 = 42
5. 72 ÷ 8 = _____
   _____ × 8 = 72
6. 32 ÷ 4 = _____
   _____ × 4 = 32

You can use multiplication to help you find the unknown values in these division questions. Complete these questions with unknown values.

7. 49 ÷ _____ = 7
   _____ × 7 = 49
8. 54 ÷ _____ = 9
   _____ × 9 = 54
9. 64 ÷ _____ = 8
   _____ × 8 = 64
10. 56 ÷ _____ = 8
    _____ × 8 = 56
11. 45 ÷ _____ = 5
    _____ × 5 = 45
12. 99 ÷ _____ = 11
    _____ × 11 = 99
13. 120 ÷ _____ = 20
    _____ × 20 = 120
14. 125 ÷ _____ = 25
    _____ × 25 = 125

SCORE /14 0-6 7-12 13-14

Quick Check

Complete these division questions.

1. 9 ÷ 9 = _____
2. 27 ÷ 3 = _____
3. 63 ÷ 21 = _____
4. 49 ÷ 7 = _____
5. 99 ÷ 9 = _____

Find the unknown values.

6. 24 ÷ _____ = 6
7. 35 ÷ _____ = 5
8. 28 ÷ _____ = 4
9. 12 ÷ _____ = 12
10. 48 ÷ _____ = 6

SCORE /10 0-4 5-8 9-10

TARGETING MATHS HOMEWORK: YEAR 5 © PASCAL PRESS ISBN: 9781925726589

Number & Algebra

AC9M5N07

Extending Division Facts

Knowing that 36 ÷ 9 = 4 helps me find 360 ÷ 9 because 360 is 10 times larger than 36.

36 ÷ 9 = 4

360 ÷ 9 = 4 × 10 = 40

Complete these division questions.

	Division	Working out
1	450 ÷ 5	45 ÷ 5 = 9 450 ÷ 5 = 9 × ___ = ___
2	720 ÷ 8	72 ÷ 8 = ___ 720 ÷ 8 = ___ × ___ = ___
3	320 ÷ 8	___ ÷ ___ = ___ _____ ÷ ___ = ___ × ___ = ___
4	720 ÷ 9	___ ÷ ___ = ___ _____ ÷ ___ = ___ × ___ = ___
5	560 ÷ 7	___ ÷ ___ = ___ _____ ÷ ___ = ___ × ___ = ___
6	810 ÷ 9	___ ÷ ___ = ___ _____ ÷ ___ = ___ × ___ = ___
7	540 ÷ 9	___ ÷ ___ = ___ _____ ÷ ___ = ___ × ___ = ___
8	420 ÷ 6	___ ÷ ___ = ___ _____ ÷ ___ = ___ × ___ = ___

SCORE /8 0-3 4-6 7-8

Statistics & Probability

AC9M5P01

Measuring Chance Events

We measure chance using a scale of 0–1, where:

- **0** means impossible
- **0·5** means an even chance
- **1** means certain.

We can show this on a chance number line, like this:

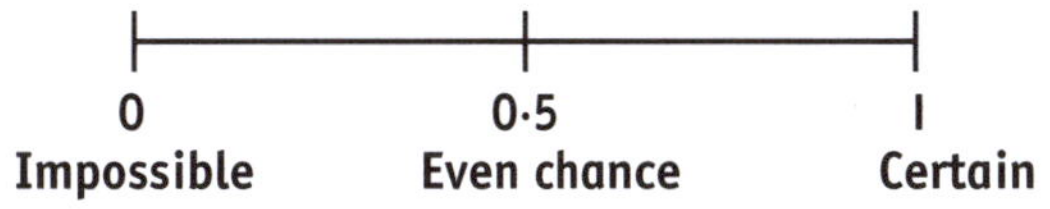

Write the letters A to E on the chance number line to show how likely it is that the following events will happen.

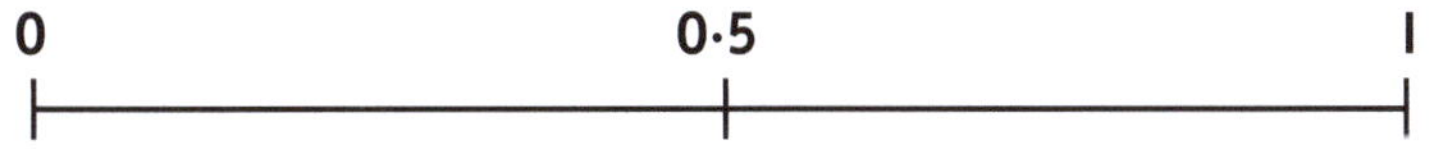

1. **A** A coin will be tossed and land on heads.
2. **B** A cow will come into the classroom.
3. **C** The stars will come out tonight.
4. **D** You will sneeze in the next hour.
5. **E** You will score 100% on this page.

SCORE /5 0-2 3-4 5

TERM 1

AC9M5M03

24-hour Time

There are 24 hours in a day. The day starts in the middle of the night, at midnight, and 12 hours later we come to midday.

Another 12 hours and we are back at midnight.

The 12 hours before midday are **am** times **(ante meridiem)** and the 12 hours after midday are **pm** times **(post meridiem)**.

On a 24-hour clock, midnight shows as 00:00 and midday shows as 12:00. The hours are numbered from midnight, so 1 o'clock in the morning is 01:00 but 1 o'clock in the afternoon is 13:00.

Complete the 24-hour digital clocks for these times.

1. Half past 3 pm [:]
2. 20 past 4 pm [:]
3. Quarter to 6 am [:]
4. Quarter past 10 am [:]
5. 35 past 9 am [:]
6. 20 past 7 in the evening [:]
7. Ten to nine in the evening [:]
8. Quarter to 8 in the morning [:]
9. 20 to four in the afternoon [:]
10. 25 to eight in the morning [:]

SCORE /10 0-4 5-8 9-10

NAPLAN*-style Questions

1. What is the unknown number in this division?

   64 ÷ _____ = 8

2. Which of these 24-hour digital clocks shows a quarter to two in the afternoon?

   ◯ 02:15 ◯ 01:45 ◯ 02:45 ◯ 13:45

3. The weather bureau has forecast a **strong chance** of rain for Monday.

   Which mark on the chance line matches their weather forecast?

   ◯ A ◯ B ◯ C ◯ D

SCORE /3 0-1 2 3

TARGETING MATHS HOMEWORK: YEAR 5 © PASCAL PRESS ISBN: 9781925726589

Calculator Magic

Leo and I have been making magic happen with our calculators.
This is how we did it:

Think of a 2-digit number between 10 and 99 and write it down.

10s	1s

Enter your number in the calculator and press **M+** to store your number in Memory.

Use this key sequence to multiply your number by 8, add the original number and divide by 9.

MR **×** **8** **+** **MR** **÷** **9** **=**   Result: _____

Use this key sequence to multiply your number by 7, add the original number and then divide by 8.

MR **×** **7** **+** **MR** **÷** **8** **=**   Result: _____

Use this key sequence to multiply your number by 6, add the original number and then divide by 7.

MR **×** **6** **+** **MR** **÷** **7** **=**   Result: _____

Over to you now!

1. Multiply by 5, add the original number and then divide by 6. Result: _____
2. Multiply by 4, add the original number and then divide by 5. Result:
3. Multiply by 3, add the original number and then divide by 4. Result: _____
4. Multiply by 2, add the original number and then divide by 3. Result: _____
5. Multiply by 1, add the original number and then divide by 2. Result: _____
6. Explain what happens each time and why that should be.

____

More Calculator Magic

- Think of a 2-digit number between 10 and 99 and write it in the chart.
- Enter the number in your calculator (e.g. 32).
- Add the reverse of your number (e.g. 23) and divide by 11.
- Write the result in the chart.
- Record your results for three different starting numbers.
  Result 1: _____ Result 2: _____ Result 3: _____
- Explain what happens each time and why that should be.

	10s	1s
Start		
Result		

____

Number & Algebra

AC9M5N08

Rounding

Numbers can be rounded up or down to the nearest 10, 100 or 1000 and beyond.

136 rounds up to 140 when rounding to the nearest 10 but rounds down to 100 when rounding to the nearest 100.

The conventions for rounding are:

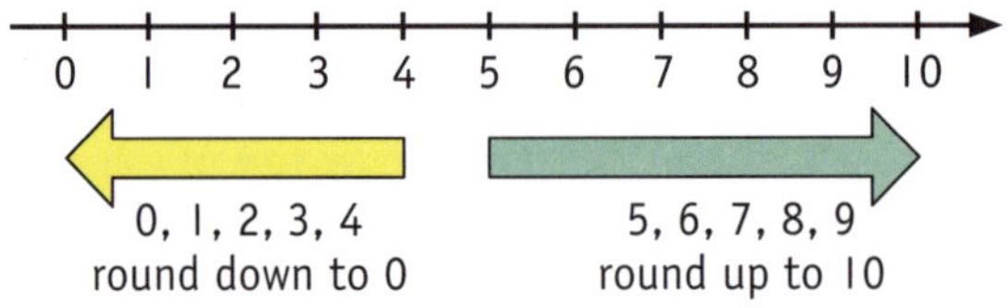

Round these numbers.

To the nearest 10

1. 153 rounds ________ to ______
2. 426 rounds ________ to ______
3. 345 rounds ________ to ______
4. 968 rounds ________ to ______

To the nearest 100

153 rounds ________ to ______

426 rounds ________ to ______

345 rounds ________ to ______

968 rounds ________ to ______

You can round decimals to the nearest whole number, tenth or hundredth.
Round these decimals.

To the nearest tenth

5. 1·363 rounds ________ to ______
6. 2·755 rounds ________ to ______
7. 4·862 rounds ________ to ______
8. 3·229 rounds ________ to ______

To the nearest hundredth

1·363 rounds ________ to ______

2·755 rounds ________ to ______

4·862 rounds ________ to ______

3·229 rounds ________ to ______

Money amounts round to the nearest 5 cents, 10 cents or whole dollar.

To the nearest 5 cents

9. $1·44 rounds ________ to ______
10. $3·57 rounds ________ to ______

To the nearest 10 cents

$1·44 rounds ________ to ______

$3·57 rounds ________ to ______

SCORE /10 0-4 5-8 9-10

Quick Check

Round these amounts to the nearest whole number or dollar.

1. 14·6 rounds ________ to ______
2. 13·25 rounds ________ to ______
3. $1·75 rounds ________ to ______
4. $2·20 rounds ________ to ______
5. $17·49 rounds ________ to ______
6. 15·376 rounds ________ to ______
7. 5·499 rounds ________ to ______
8. $38·50 rounds ________ to ______
9. $5·95 rounds ________ to ______
10. $9·51 rounds ________ to ______

SCORE /10 0-4 5-8 9-10

TARGETING MATHS HOMEWORK: YEAR 5 © PASCAL PRESS ISBN: 9781925726589

Number & Algebra

AC9M5N09

Round and Adjust

Round and adjust is a strategy that I first learned about when I was doing Natural Maths. It makes adding large numbers so easy!

For 57 + 34, round 57 up to 60 by adding 3. Then adjust by taking 3 off the 34.

I write it like this:

$57^{+3} + 34^{-3}$
= 60 + 31 and now we can add 31 to a friendly number.
= 91

Complete these examples.

1. $78^{+\_\_} + 63^{-\_\_}$
   = ______ + ______
   = ______

2. $37^{+\_\_} + 89^{-\_\_}$
   = ______ + ______
   = ______

3. $43^{-\_\_} + 69^{+\_\_}$
   = ______ + ______
   = ______

Round and adjust works for money amounts as well.

For \$6·85 + \$3·45, round \$6·85 up to \$7 by adding 15c. Then adjust by taking 15c from \$3·45.

$\$6{\cdot}85^{+15c} + \$3{\cdot}45^{-15c}$
= \$7 + \$3·30 and now we can add \$3·30 to a whole-dollar amount.
= \$10·30

Complete these examples.

4. $\$7{\cdot}75^{+\_\_\_} + \$3{\cdot}45^{-\_\_\_}$
   = ________ + ________
   = ________

5. $\$16{\cdot}89^{+\_\_\_} + \$3{\cdot}25^{-\_\_\_}$
   = ________ + ________
   = ________

6. $\$27{\cdot}35^{-\_\_\_} + \$12{\cdot}80^{+\_\_\_}$
   = ________ + ________
   = ________

SCORE /6 0-2 3-4 5-6

Statistics & Probability

AC9M5ST01

The Bean Growth Science Experiment

The line graph shows the heights of two beans. Bean A was watered every day and Bean B was watered every second day.

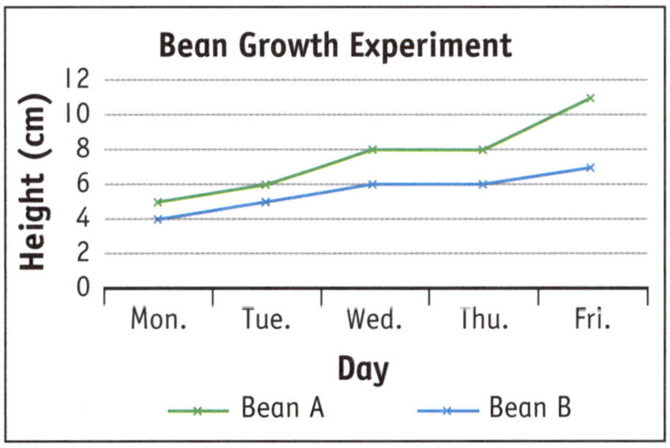

1. Which bean grew more and by how much?
   ____
2. Between which two days did neither bean grow?
   ____
3. Between which two days did Bean A grow most?
   ____
4. How much did Bean B grow between Monday and Wednesday? ___ cm
5. How much taller than Bean B was Bean A on Friday? ___ cm

SCORE /5 0-2 3-4 5

AC9M5SP02

Draw Me a House

Follow the coordinates to draw the outline of a house on the grid.

1. Label the axes of the grid from 0 to 10.
2. Draw from (1, 1) to (1, 3).
3. Draw from (1, 3) to (4, 6).
4. Draw from (4, 6) to (7, 4).
5. Draw from (7, 4) to (7, 1).
6. Draw from (7, 1) to (1, 1).
7. Join the points (2, 2), (2, 3), (3, 3), (3, 2) and back to (2, 2).
8. Add a front door and write the coordinates.
   Join (__, __), (__, __), (__, __) and (__, __)
9. Add a lean-to shed on the right-hand side of the house and write the coordinates for that.
   Join (__, __), (__, __), (__, __) and (__, __).

SCORE /9 0-4 5-7 8-9

NAPLAN*-style Questions

1. Which amount is the best estimate for $3·95 + $7·80?
   ◯ $10 ◯ $11 ◯ $12 ◯ $13

2. On Tuesday, Miro measured the temperature every 2 hours from 10:00 am.

Time	10:00	12:00	14:00	16:00
Temperature (°C)	12	18	22	17

   Which of these charts represents Miro's data?

◯ 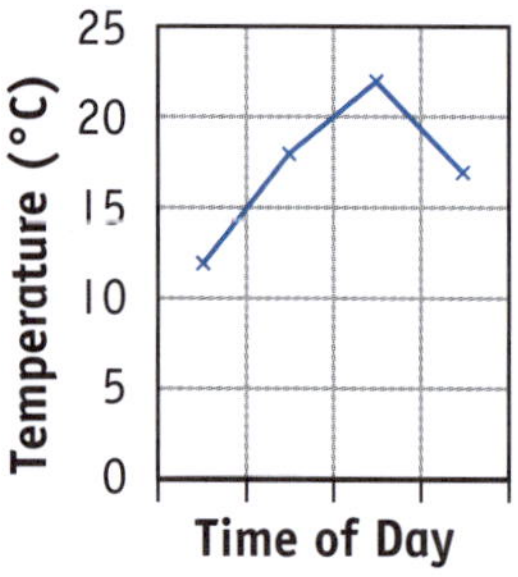

◯ 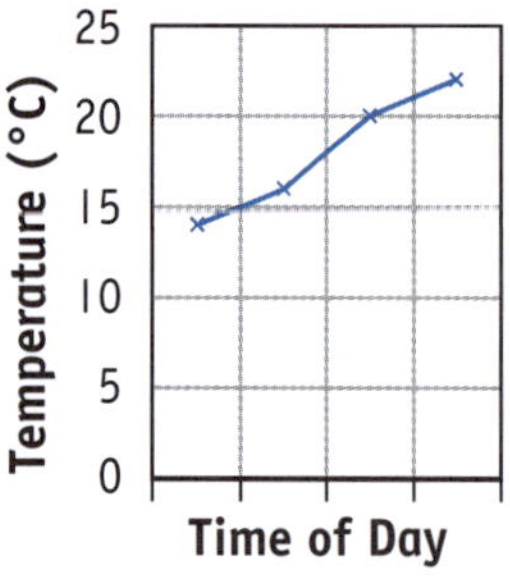

◯ 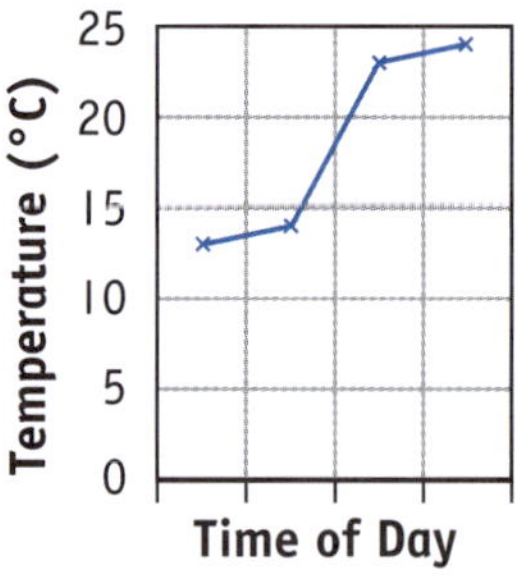

◯ 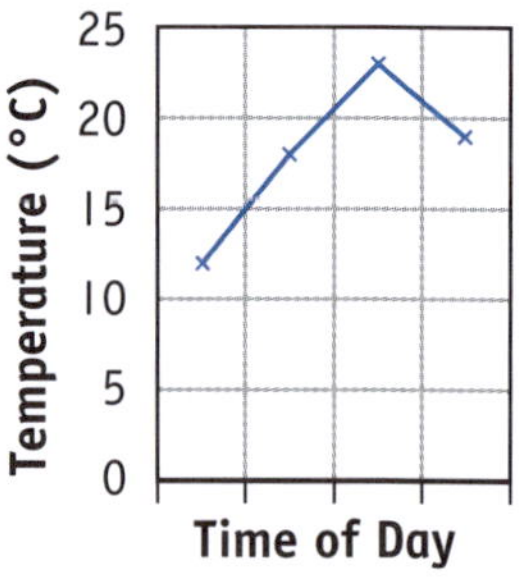

3. What are the coordinates of the × on this grid?
   ◯ (2, 3) ◯ (3, 2)
   ◯ (2, 4) ◯ (0, 2)

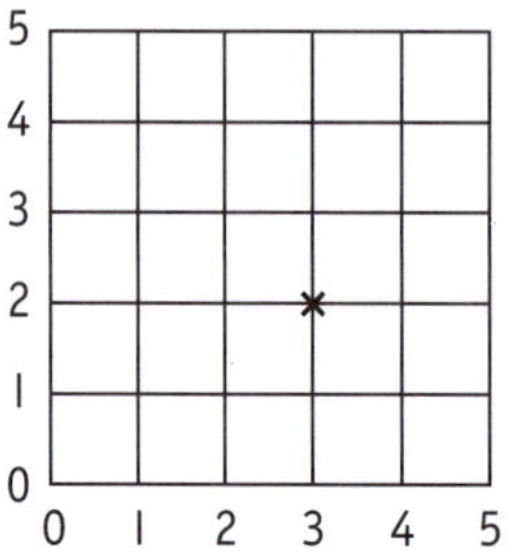

SCORE /3 0-1 2 3

TARGETING MATHS HOMEWORK: YEAR 5 © PASCAL PRESS ISBN: 9781925726589

Problem of the Week

AC9M5ST01

Training Data

Ella, Will, Asha and Ravi are in the basketball team. They were practising shooting hoops last week and this is a graph of the results of their training.

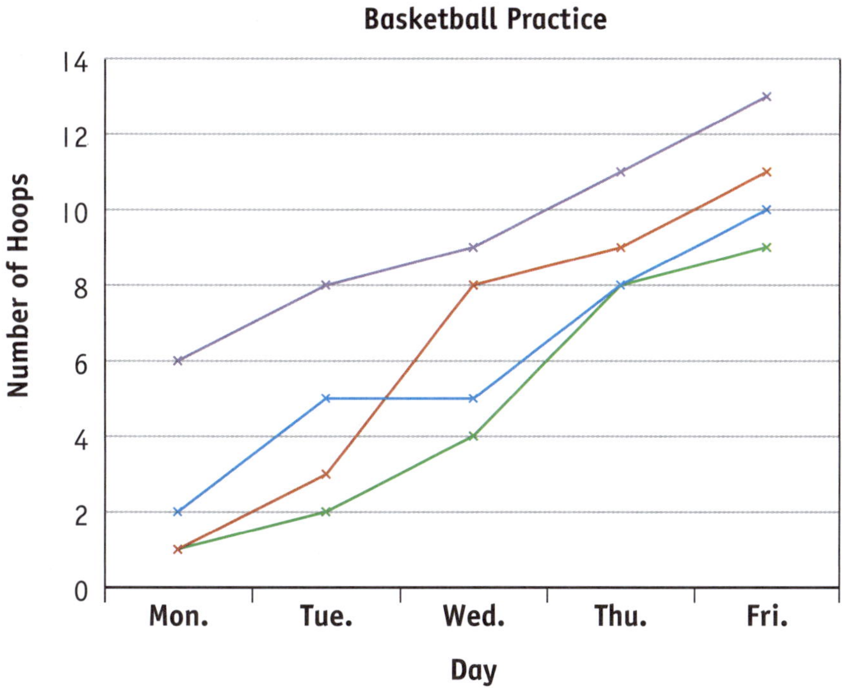

Which colour line shows the daily shooting scores for each team member?
Here are some clues:

I had a plateau – there was no change in my scores from one day to the next.
Asha's data is shown in the colour ____________.

I was ahead every day of the week.
Ravi's data is shown in the colour ____________.

I made the most progress over the week.
Ella's data is shown in the colour ____________.

From Monday to Thursday, I improved by 7 hoops.
Will's data is shown in the colour ____________.

By the end of next week's training, who do you think will be shooting the most hoops?

____

Why do you think that?

____

Number & Algebra

Remainders

Sometimes a division doesn't work out exactly. For example, 9 ÷ 4 is 2 with 1 left over.
The 2 gives the number of 4s that you can make from 9 and is called the **quotient**.
The left-over part is called the **remainder**.
We write this as:
9 ÷ 4 = 2 r 1, where **r** stands for **remainder**.

Write the quotients and remainders for these divisions.

1. 12 ÷ 5 = ___ r ___
2. 17 ÷ 4 = ___ r ___
3. 23 ÷ 7 = ___ r ___
4. 33 ÷ 8 = ___ r ___
5. 39 ÷ 5 = ___ r ___
6. 56 ÷ 9 = ___ r ___

Remainders with money

It is a little bit different with money because we can also divide the remainder.
For example, $9 ÷ 4 is $2 with $1 left over.
So we have to find: $1 ÷ 4 = 25c
So **$9 ÷ 4 = $2 and 25c = $2·25c**.

Complete these money divisions.

7. $13 ÷ 4 = ____ r ____
   $____ ÷ 4 = ____ cents
   $13 ÷ 4 = $______
8. $15 ÷ 2 = ____ r ____
   $____ ÷ ____ = ____ cents
   $____ ÷ ____ = $______
9. $16 ÷ 5 = ____ r ____
   $____ ÷ ____ = ____ cents
   $____ ÷ ____ = $______
10. $26 ÷ 4 = ____ r ____
   $____ ÷ ____ = ____ cents
   $____ ÷ ____ = $______

SCORE /10 (0-4) (5-8) (9-10)

Quick Check

Write the quotients and remainders for these divisions.

1. 26 ÷ 5 = ____ r ____
2. 33 ÷ 6 = ____ r ____
3. 19 ÷ 4 = ____ r ____
4. 28 ÷ 5 = ____ r ____
5. 101 ÷ 10 = ____ r ____

Complete these money divisions.

6. $13 ÷ 2 = $______
7. $33 ÷ 6 = $______
8. $19 ÷ 4 = $______
9. $28 ÷ 5 = $______
10. $101 ÷ 10 = $______

SCORE /10 (0-4) (5-8) (9-10)

TARGETING MATHS HOMEWORK: YEAR 5 © PASCAL PRESS ISBN: 9781925726589

Number & Algebra

UNIT 8

AC9M5N09

TERM 1

Division Word Problems

Write the number sentences and answers for these division questions.

1. Jade and Leo have $9 to share equally between them.
How much should they each get?

_______ ÷ _______ = _______

2. A bag of 4 apples costs $6.
How much does each apple cost?

_______ ÷ _______ = _______

3. The neighbour gave Miro $32 to walk the dog for five days next week.
How much is that for each day?

_______ ÷ _______ = _______

4. If Gran buys new shoes for Ella for $122 and Ella pays her back over four weeks,
how much will Ella pay Gran each week?

_______ ÷ _______ = _______

5. 5 litres of fertiliser for the lawn costs $27.
How much is that per litre?

_______ ÷ _______ = _______

SCORE /5 0-2 3-4 5

Statistics & Probability

AC9M5ST03

At the Movies

This chart shows the number of people in the cinema watching the latest movie.

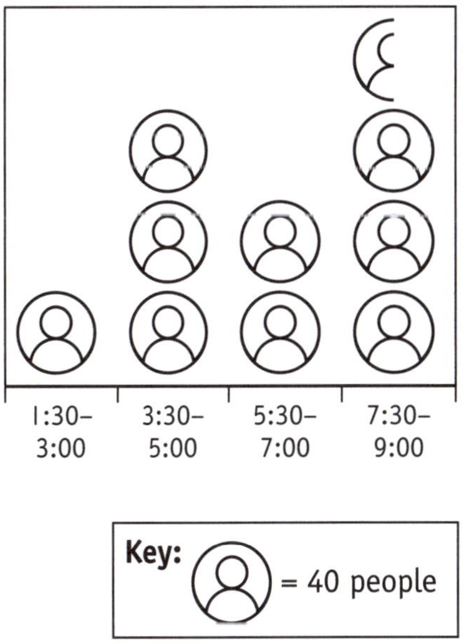

1. How many more people were at the 3:30 show than the 1:30 show?

____

2. How many fewer people were at the 1:30 show than at the 7:30 show?

____

3. How many people altogether saw the movie that day?

____

4. There was a 9:30 show which had more people than the 3:30 show. How many people might have been at the 9:30 show?

____

5. Why do you think the later shows had more people attending than the earlier shows?

____

SCORE /5 0-2 3-4 5

TERM 1

AC9M5M04

Measuring and Constructing Angles

You will need a 360° protractor for this activity. This is what mine looks like:

Measure these angles and say if they are acute, right, obtuse, straight or reflex.

1. _____ ° and ___________

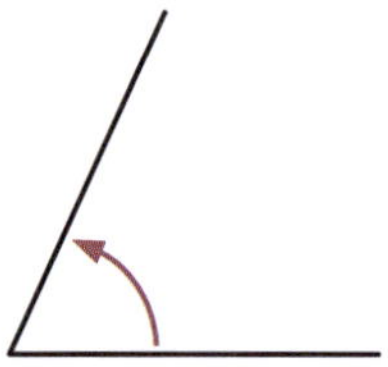

2. _____ ° and ___________

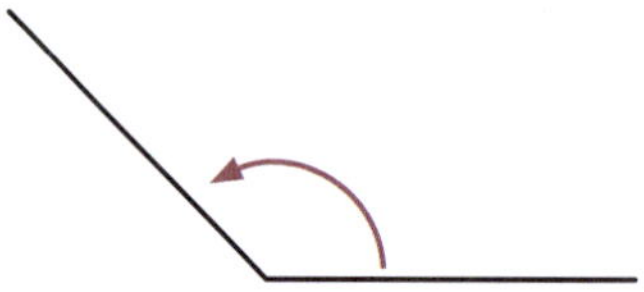

3. _____ ° and ___________

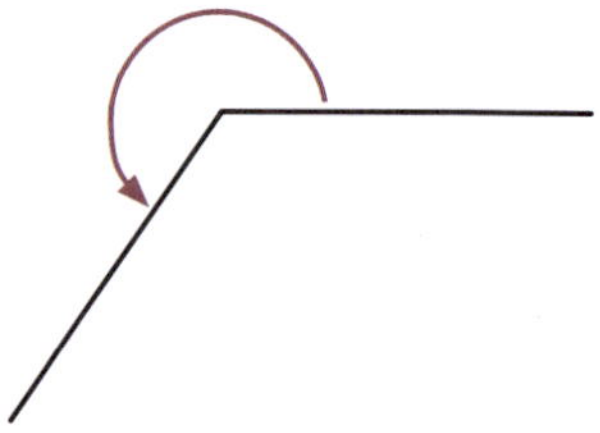

Use your protractor to draw these angles.

4. 45° angle	5. 120° angle	6. 280° angle

SCORE /6 0-2 3-4 5-6

NAPLAN*-style Questions

1. The gardener planted 24 seedlings into 4 equal rows.
   How many seedlings did she plant in each row?
   ◯ 28 ◯ 20 ◯ 4 ◯ 6

2. The graph shows the number of students competing in three sports events.
   How many students competed in these events altogether?
   ☐ students

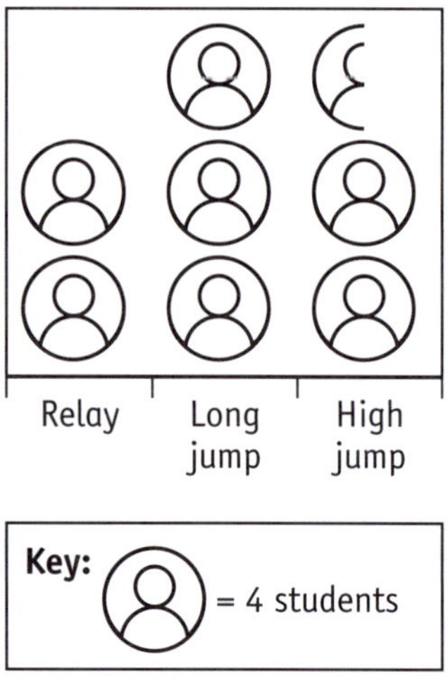

3. A bicycle ramp is supposed to be 30°.
   Which of these ramps is closest to that?

 ◯

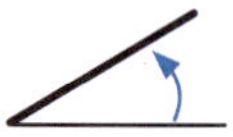 ◯

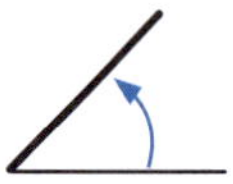 ◯

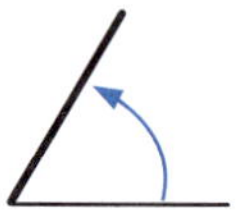 ◯

SCORE /3 0-1 2 3

TARGETING MATHS HOMEWORK: YEAR 5 © PASCAL PRESS ISBN: 9781925726589

Problem of the Week

AC9M5N07

Magic Hat Tricks

A famous mathemagician performs magic hat tricks with rabbits.
He often uses different hats, and he has lots of rabbits.

People always ask him how many rabbits he has.
Being a mathemagician, he always gives a trick answer.

Here is one of them:

**"When there are two rabbits in each hat, one rabbit has to stay in the cage.
But when there are three rabbits in each hat, 2 empty hats are left in the cupboard."**

How many rabbits and how many hats does the mathemagician have?

I supposed that the mathemagician had 1, 2, 3 hats and then worked out how many rabbits he would have with the first clue.

I used this table:

Hats	1	2	3	4	5	6	7	8
Rabbits								

Then I made the same table with the second clue:

Hats	1	2	3	4	5	6	7	8
Rabbits								

Try using Leo's method to solve the hats and rabbits problem.

The mathemagician has ____ hats and ____ rabbits.

The mathemagician also performs hat tricks with birds.
This time his clues are:

**"When there are three birds in each hat, two birds have to stay in the cage.
But when there are four birds in each hat, three empty hats are left in the cupboard."**

How many birds and how many bird-hats does the mathemagician have?
Draw the tables that will help you solve the problem:

The mathemagician has ____ bird-hats and ____ birds.

TERM 1 REVIEW

Review: Unit 1 and Unit 2

AC9M5N01, AC9M5N02, AC9M5M01, AC9M5ST01

TERM 1

Write these fractions in decimal form.

(1) $\frac{3}{10}$ = __·____

(2) $\frac{14}{100}$ = __·____

(3) $\frac{5}{100}$ = __·____

(4) $\frac{6}{1000}$ = __·____

(5) $\frac{13}{1000}$ = __·____

(6) $\frac{16}{100}$ = __·____

Complete these decimal additions.

(7) 0·9 + 0·6 = $\frac{}{10}$ + $\frac{}{10}$ = $\frac{}{10}$ or __·____

(8) 0·03 + 0·08 = —— + —— = —— or __·____

(9) 0·004 + 0·005 = —— + —— = —— or __·____

(10) 0·008 + 0·007 = —— + —— = —— or __·____

Match the item to be measured with the unit by which it would be measured.

(11) Area of a rug	**A** kg
(12) Capacity of a jug	**B** m^2
(13) Mass of a sack of potatoes	**C** cm
(14) Width of this page	**D** mL
(15) Mass of a blueberry	**E** g

Write the factors of these numbers.

(16) The factors of 5 are ________________.

(17) The factors of 12 are ________________.

(18) The factors of 13 are ________________.

(19) The factors of 16 are ________________.

Six people gave ratings of three take-away meals. They used a scale of 1–5, where 1 was the least liked and 5 was the most liked.

Take-away	Voters					
	A	**B**	**C**	**D**	**E**	**F**
Fish and Chips	3	3	2	2	1	1
Pizza	4	5	3	1	5	4
Burger	4	5	4	3	4	3

(20) Which take-away meal got the lowest rating? ____________

(21) Which take-away meal got the highest rating? ____________

(22) Which voter liked take-away meals the least? ____

(23) Which voter liked take-away meals the most? ____

(24) How much higher was the rating for Burger than for Fish and Chips? ____

SCORE /24 0-11 12-22 23-24

TARGETING MATHS HOMEWORK: YEAR 5 © PASCAL PRESS ISBN: 9781925726589

Review: Unit 3 and Unit 4

AC9M5N04, AC9M5N07, AC9M5N08, AC9M5M02, AC9M5P01

Convert the mixed numbers into improper fractions.

1. $1\frac{3}{5} = \frac{\quad}{\quad}$
2. $3\frac{1}{4} = \frac{\quad}{\quad}$
3. $1\frac{3}{10} = \frac{\quad}{\quad}$
4. $2\frac{5}{8} = \frac{\quad}{\quad}$
5. $2\frac{1}{2} = \frac{\quad}{\quad}$
6. $4\frac{2}{3} = \frac{\quad}{\quad}$

Place these fractions on the number line.

7. $\frac{3}{8}$, $1\frac{1}{2}$, $\frac{5}{4}$, $1\frac{1}{8}$

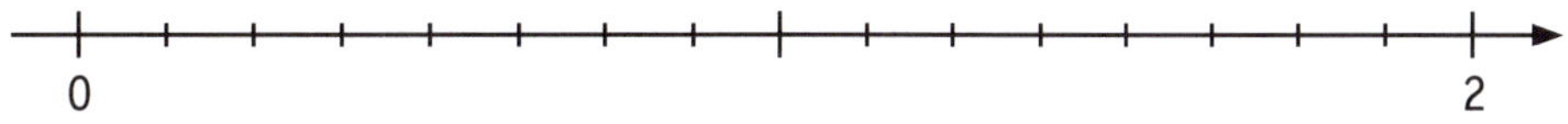

Circle the largest fraction in each group.

8. $1\frac{3}{4}$, $\frac{11}{8}$, $1\frac{1}{2}$
9. $\frac{8}{10}$, $\frac{11}{10}$, $1\frac{1}{5}$

Convert these fractions and then add them

10. $\frac{1}{6} + \frac{1}{3} + \frac{1}{2} = \frac{\quad}{\quad} + \frac{\quad}{\quad} + \frac{\quad}{\quad} =$
11. $\frac{1}{4} + \frac{3}{8} + \frac{3}{4} = \frac{\quad}{\quad} + \frac{\quad}{\quad} + \frac{\quad}{\quad} =$

12. Which spinner has the better chance of spinning a 2?

Write these fractions as decimals and percentages.

13. $\frac{6}{100}$ is the same as 0·_____ and _____%.
14. $\frac{24}{100}$ is the same as 0·_____ and _____%.
15. $\frac{7}{10}$ is the same as 0·_____ and _____%.

What would a $100 pair of shoes cost with these discounts?

16. After a 20% discount, the shoes would cost $_____.
17. After a 15% discount, the shoes would cost $_____.
18. After an 8% discount, the shoes would cost $_____.

Write the area and perimeter of these shapes that have been made with 1 m squares.

19. 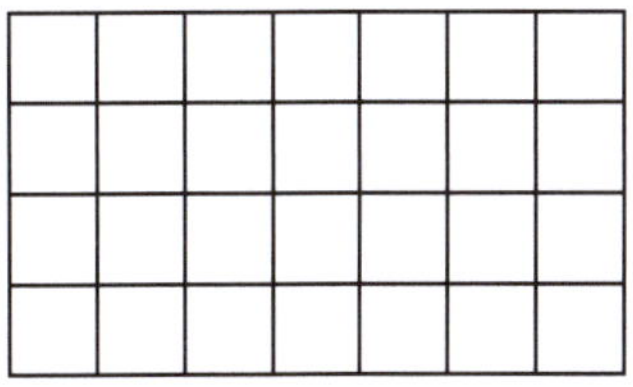

Area = _____ m^2 Perimeter = _____ m

20. 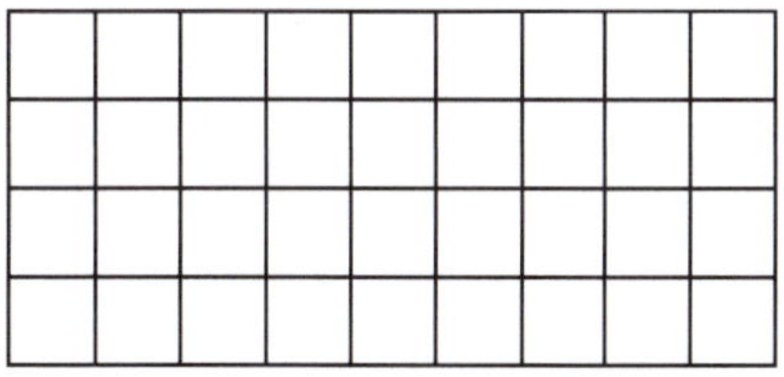

Area = _____ m^2 Perimeter = _____ m

SCORE /20 0-9 10-18 19-20

TERM 1 REVIEW

Review: Unit 5 and Unit 6

AC9M5N06, AC9M5N07, AC9M5N08, AC9M5N09, AC9M5A02, AC9M5M03, AC9M5SP02, AC9M5P02

TERM 1

Complete these multiplications.

(1) 56 × 5

____ + ___ = ____

× __ [|] = ______

(2) 83 × 7

____ + ___ = ____

× __ [|] = ______

Write reasonable estimates for these multiplications.

(3) 47 × 6

Between ______ and ______

(4) 73 × 7

Between ______ and ______

Complete these money multiplications.

(5) 6 × $7·85

$___ + ___c + ___c = $_______

× __ [| |] = $_______

(6) 4 × $6·38

$___ + ___c + ___c = $_______

× __ [| |] = $_______

This table shows how many heads and tails Asha tossed in 8 games:

<table>
<tr><td>Heads</td><td>|||</td></tr>
<tr><td>Tails</td><td>卌</td></tr>
</table>

Asha's results were:

(7) ___ heads out of ___ tosses.

(8) ___ tails out of ____ tosses.

List the coordinates of the corners of the hexagon.

(9)

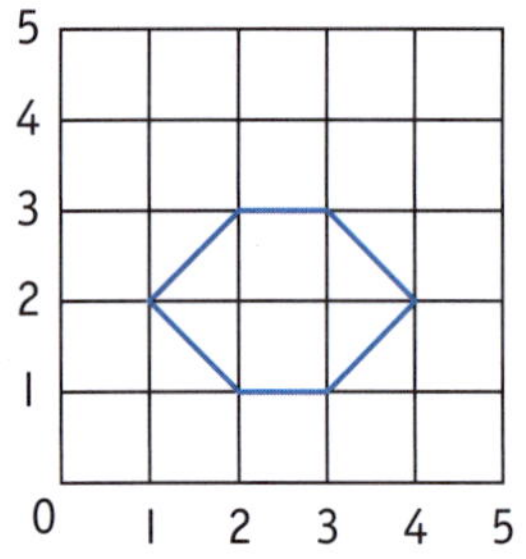

(__,__) (__,__) (__,__)

(__,__) (__,__) (__,__)

Write the related division and multiplication facts.

(10) 45 ÷ 9 = ___

___ × 9 = ___

(11) 72 ÷ 8 = ___

___ × 8 = ___

Complete these division questions.

(12) 320 ÷ 8 ___ ÷ ___ = ___

_____ ÷ ___ = ___ × ___ = ___

(13) 540 ÷ 9 ___ ÷ ___ = ___

_____ ÷ ___ = ___ × ___ = ___

Complete the digital clocks using 24-hour time.

(14) Ten to eleven am

(15) Quarter to seven pm

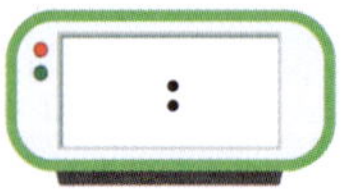

(16) 5 minutes to midnight

SCORE /16 0-7 8-14 15-16

TARGETING MATHS HOMEWORK: YEAR 5 © PASCAL PRESS ISBN: 9781925726589

Review: Unit 7 and Unit 8

TERM 1 REVIEW

AC9M5N07, AC9M5N08, AC9M5N09, AC9M5M04, AC9M5ST01, AC9M5ST03

TERM 1

Round these numbers to the nearest tenth.

(1) 3·75 ______ (2) 2·84 ______ (3) 3·96 ______

Round these numbers to the nearest hundredth.

(4) 1·768 ______ (5) 1·345 ______ (6) 1·272 ______

Round these amounts to the nearest 10c.

(7) $1·73 ______ (8) $3·68 ______

Round these amounts to the nearest dollar.

(9) $2·55 ______ (10) $3·46 ______

Use round and adjust for these additions.

(11) $77^{+\_\_} + 26^{-\_\_}$
= ______ + ______
= ______

(12) $56^{+\_\_} + 75^{-\_\_}$
= ______ + ______
= ______

(13) $42^{-\_\_} + 58^{+\_\_}$
= ______ + ______
= ______

The graph at the right shows a kitten's growth.

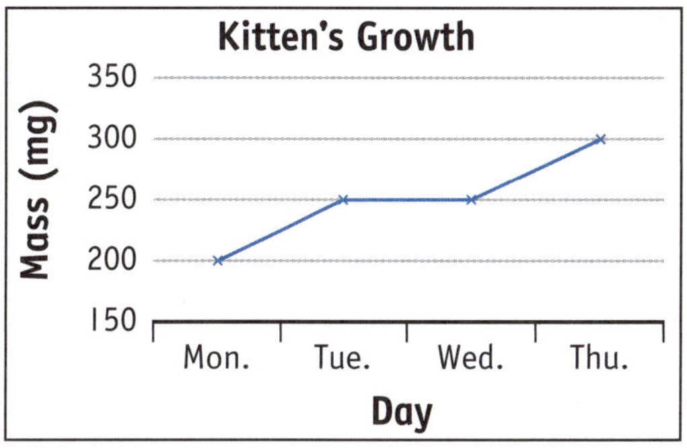

(14) How much weight did the kitten gain between Tuesday and Thursday?

______ mg

(15) How heavy was the kitten on Wednesday?

______ mg

Complete these division and remainder questions.

(16) 14 ÷ 4 = ___ r ___ (18) $21 ÷ 4 = $________

(17) 27 ÷ 5 = ___ r ___ (19) $17 ÷ 5 = $________

The graph at the right shows the crowd sizes at junior football games.

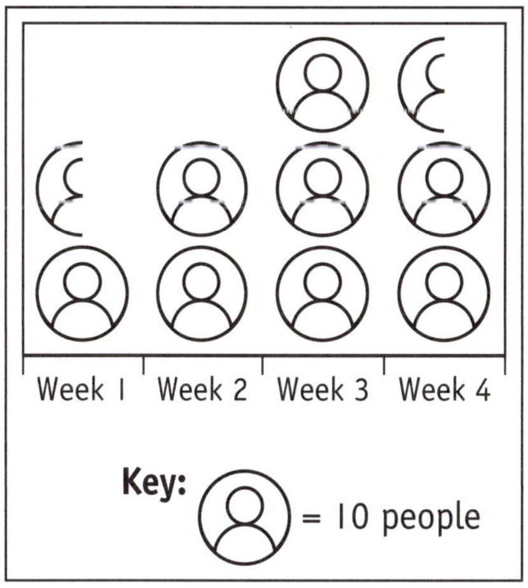

(20) How many people altogether attended the four football games?

____

(21) How many more people attended the Week 3 game than the Week 1 game?

____

Measure these angles and write what type they are.

(22) _____ ° and ___________

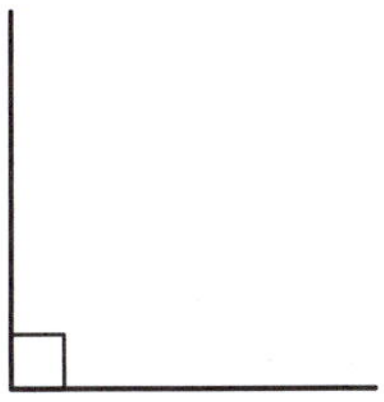

(23) _____ ° and ___________

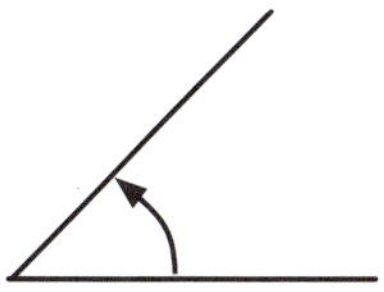

(24) _____ ° and ___________

SCORE /24 0-11 12-22 23-24

TARGETING MATHS HOMEWORK: YEAR 5 © PASCAL PRESS ISBN: 9781925726589

TERM 2

Finding Multiples

When you multiply two numbers together, the result is called a **multiple**.
Because the two numbers divide evenly into the multiple, they are called **factors** of that multiple. For example:

6 is the **multiple** of 2 and 3 because 6 = 2 × 3 and 2 and 3 are **factors** of 6.

We can also find multiples of more than two factors:

2 × 3 × 4 = 24 which is a **multiple** of 2, 3 and 4 and
2, 3 and 4 are **factors** of 24.

Find the multiple of each set of factors.

1. Factors 2, 5, 6. Multiple ____
2. Factors 1, 3, 4. Multiple ____
3. Factors 3, 4, 5. Multiple ____
4. Factors 2, 3, 5. Multiple ____
5. Factors 6, 5, 10. Multiple ____
6. Factors 6, 4, 3. Multiple ____
7. Factors 3, 7, 4. Multiple ____
8. Factors 5, 2, 8. Multiple ____
9. Factors 2, 6, 4. Multiple ____
10. Factors 4, 3, 6. Multiple ____

Finding the Lowest Common Multiple

The numbers 2, 5 and 6 are all factors of 60, but 60 is not the smallest number that is a multiple of 2, 5 and 6.

30 is the smallest number that is a multiple of 2, 5 and 6.
We call 30 the **lowest common multiple** of 2, 5 and 6. For example:

3, 4 and 6 are all factors of 72, but 72 is not the smallest multiple of 3, 4 and 6.
The **lowest common multiple** of 3, 4 and 6 is **12**.

Find the lowest common multiple (also called the LCM) of each set of factors.

11. Factors 5, 2, 6. LCM ____
12. Factors 3, 4, 9. LCM ____
13. Factors 2, 3, 7. LCM ____
14. Factors 6, 3, 4. LCM ____
15. Factors 8, 2, 6. LCM ____
16. Factors 8, 2, 3. LCM ____
17. Factors 4, 5, 2. LCM ____
18. Factors 3, 6, 9. LCM ____
19. Factors 3, 6, 7. LCM ____
20. Factors 8, 3, 9. LCM ____

SCORE /20 0-9 10-18 19-20

Quick Check

Write the lowest common multiple.

1. 5, 2, 4 ____
2. 13, 2, 3 ____
3. 3, 1, 6 ____
4. 26, 1, 3 ____
5. 1, 2, 3, 4 ____
6. 3, 6, 2 ____
7. 3, 4, 8 ____
8. 5, 3, 10 ____
9. 7, 2, 14 ____
10. 5, 12, 15 ____

SCORE /10 0-4 5-8 9-10

TARGETING MATHS HOMEWORK: YEAR 5 © PASCAL PRESS ISBN: 9781925726589

Number & Algebra

AC9M5N01

Decimals on a Number Line

Remember!

Between 0 and 1 there are tenths.
Between 0·0 and 0·1 there are hundredths.
Between 0·00 and 0·01 there are thousandths.
So the ↓ on this number line represents 0·025:

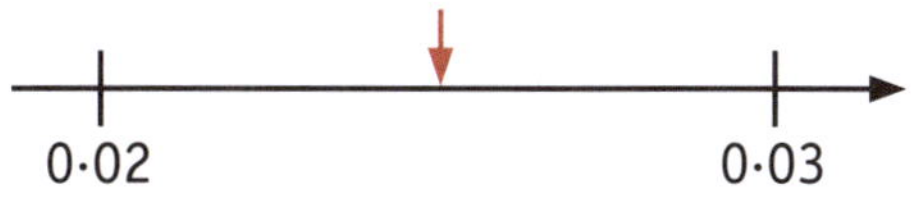

Write the decimals marked by the red arrow on these number lines.

1. 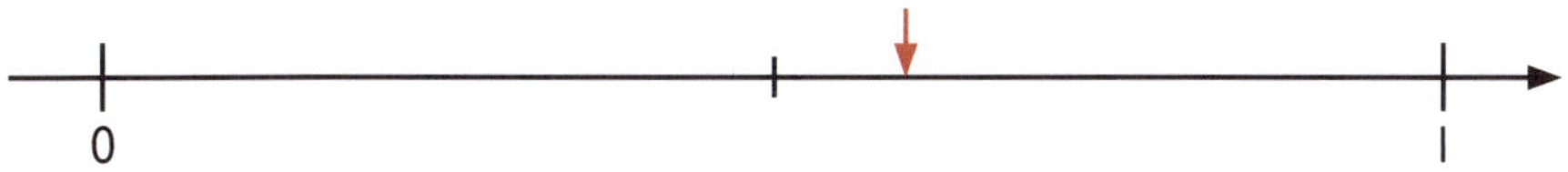

2. 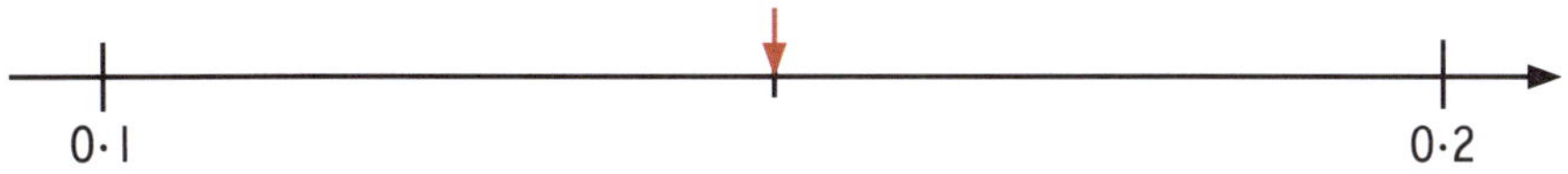

3. 

4. 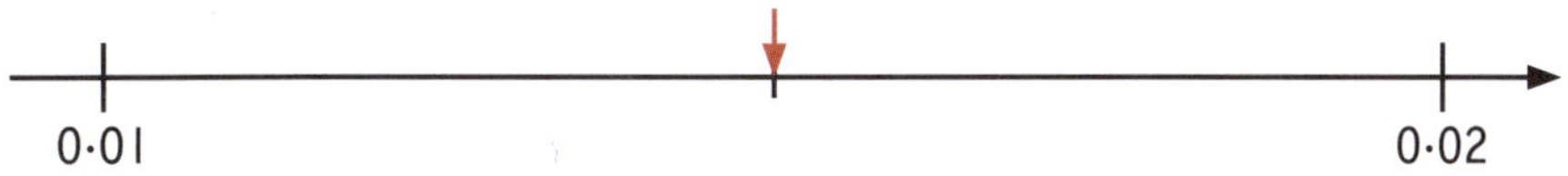

5. 

6. 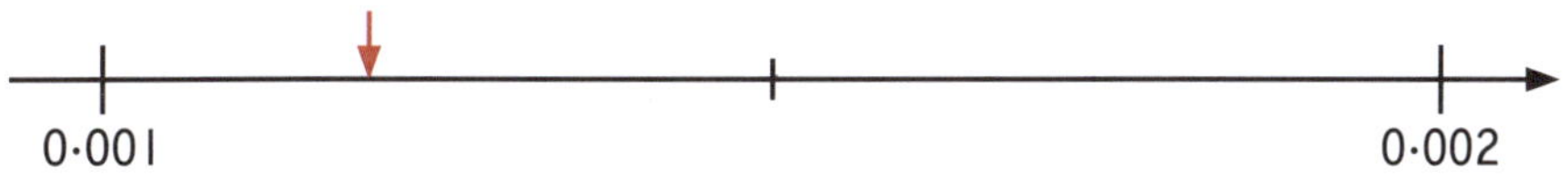

You can use a ruler to help you work out which decimal to write.

SCORE /6 0-2 3-4 5-6

Statistics & Probability

AC9M5P01

Dice Totals

You can use two dice to make all the totals from 2 to 12.

1. Complete this table of the totals that dice pairs make.
2. Which totals are you least likely to throw?

   ___ and ___

3. Which total has the best chance of being thrown?

   ____

4. If you throw two dice now, which total do you think you would make?

   ____

5. Why? ______________________________

		Dice 1					
		1	2	3	4	5	6
Dice 2	1						
	2						
	3						
	4						
	5						
	6						

SCORE /5 0-2 3-4 5

AC9M5M01

Scale it Up

Use the clues to enlarge the shape in the box below.

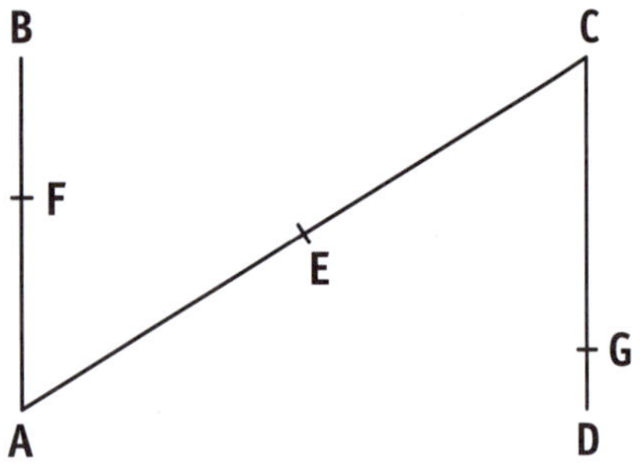

1. **A** to **B** = 60 mm
2. **A** to **C** = 11·6 cm and **C** is directly to the right of **B**.
3. **C** to **D** = 6 cm and **D** is directly below **C**.
4. **E** is halfway between **A** and **C**.
5. **F** is 24 mm from **B**.
6. **G** is 10 mm from **D**.

SCORE /6 0-2 3-4 5-6

NAPLAN*-style Questions

1. What is the number marked × on this decimal number line?

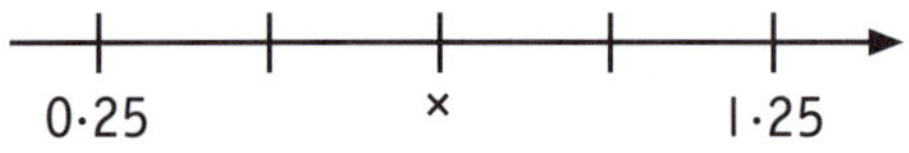

- 0·4
- 0·5
- 0·75
- 1

2. Ravi cut a quarter of a metre of elastic. How many millimetres did he cut?

- 50 mm
- 25 mm
- 500 mm
- 250 mm

3. Sam made this green and white spinner. What is the chance of spinning white?

- $\frac{1}{2}$
- $\frac{1}{4}$
- $\frac{1}{3}$
- $\frac{2}{3}$

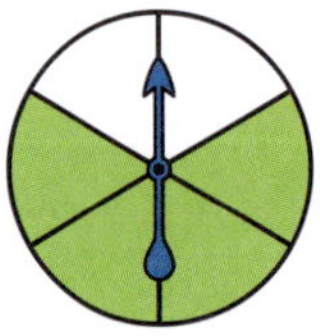

SCORE /3 0-1 2 3

TARGETING MATHS HOMEWORK: YEAR 5 © PASCAL PRESS ISBN: 9781925726589

AC9M5P01

Two-dice Bingo

We have been playing Two-Dice Bingo using bingo cards that we made for ourselves.

We made the rule that you can repeat totals on your board, but when you roll that total, you can only cross it off once.

Here are our boards after three rounds of the game:

Ella

7	~~5~~	~~7~~
~~6~~	7	4

Will

~~5~~	6	7
8	~~7~~	~~6~~

Asha

2	4	~~6~~
3	~~5~~	~~7~~

Ravi

2	~~7~~	~~6~~
9	6	~~5~~

1. Who do you think has the best chance of winning? ____________
2. Why do you think that?

____

3. Here are the next throws. Mark the results on the boards to find out who won.

Round 4

Round 6

Round 8

Round 5

Round 7

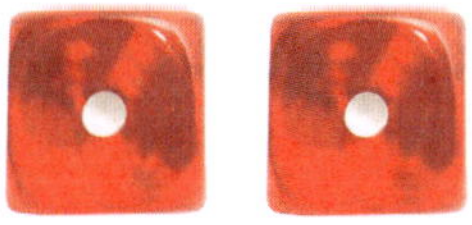

Round 9

4. Are you surprised by the winner? ______
5. Why or why not? ______________________________________
6. Make up your own winning bingo card.

7. Why do you think it will be a winner?

____

Truth Sentences

"Good morning class. When I was at school, we did **truth sentences** in our mathematics lessons. We had to read an equation and say whether the two parts were **equal (=)** or **not equal (≠)**.

I really enjoyed truth sentences and would like to share some with you.

Here are two examples:

6 × 4 ☐ 6 × 2 × 2

The answer is 6 × 4 = 6 × 2 × 2.

24 ÷ 6 ☐ 6 ÷ 24

The answer is 24 ÷ 6 ≠ 6 ÷ 24.

Here are truth sentences for you to solve.
Write = or ≠ in the box to show your answer.

1. 24 ÷ 8 ☐ 24 ÷ 4 ÷ 4
2. 24 × 8 ☐ 24 × 4 × 2
3. 36 ÷ 9 ☐ 36 ÷ 3 ÷ 3
4. 6 × 9 ☐ 3 × 6 + 3 × 6
5. 16 × 6 ☐ 16 × 3 × 2
6. 15 ÷ 3 ☐ 30 ÷ 6
7. 27 ÷ 9 ☐ 9 × 3
8. 28 ÷ 7 ☐ 28 ÷ 3 ÷ 4
9. 6 × 4 ☐ 12 × 2
10. 24 × 3 ☐ 12 × 3 × 2
11. 16 × 6 ☐ 96 ÷ 2 ÷ 3
12. 12 ÷ 6 ☐ 6 × 2
13. 12 × 6 ☐ 12 × 3 × 3
14. 4 × 3 ☐ 12 ÷ 3
15. 12 ÷ 4 ☐ 12 ÷ 3 ÷ 1
16. 3 × 4 ☐ 6 × 2
17. 42 ÷ 6 ☐ 28 ÷ 4
18. 45 ÷ 9 ☐ 45 ÷ 3 ÷ 3
19. 13 × 4 ☐ 10 × 3 × 4
20. 16 ÷ 4 ☐ 8 ÷ 2

SCORE /20 0-9 10-18 19-20

Quick Check

Draw lines to match the divisions with their related multiplications.

	Division	Multiplication
1	12 ÷ 4	9 × 4
2	16 ÷ 8	4 × 7
3	28 ÷ 4	3 × 4
4	36 ÷ 9	7 × 8
5	56 ÷ 7	2 × 8

	Division	Multiplication
6	81 ÷ 9	9 × 7
7	63 ÷ 7	8 × 7
8	56 ÷ 8	9 × 12
9	33 ÷ 11	9 × 9
10	108 ÷ 9	11 × 3

SCORE /10 0-4 5-8 9-10

TARGETING MATHS HOMEWORK: YEAR 5 © PASCAL PRESS ISBN: 9781925726589

Using Factors for Mental Multiplication

I like to use factors to help with multiplication.

This is how it works. 6 has 2 and 3 as factors so to find 26 × 6, you can double 26 to make 52 and then multiply 52 by 3 to get the answer, 156.

Here are two examples:

To find 16 × 4, use 4 = 2 × 2.
The first step is to find 16 × 2 = 32 and the second step is 32 × 2 to get 64.

To find 16 × 6, use 6 = 2 × 3.
The first step is to find 16 × 2 = 32 and the second step is 32 × 3 to get 96.

Use factors to solve these multiplications.

	First step	Second step
(1) For 23 × 4, use 4 = 2 × 2	____ × ___ = _____	____ × ___ = _____
(2) For 23 × 6, use 6 = 2 × 3	____ × ___ = _____	____ × ___ = _____
(3) For 23 × 8, use 8 = 4 × 2	____ × ___ = _____	____ × ___ = _____
(4) For 23 × 9, use 9 = ___ × ___	____ × ___ = _____	____ × ___ = _____
(5) For 32 × 4, use ___ = ___ × ___	____ × ___ = _____	____ × ___ = _____
(6) For 32 × 6, use ___ = ___ × ___	____ × ___ = _____	____ × ___ = _____
(7) For 32 × 9, use ___ = ___ × ___	____ × ___ = _____	____ × ___ = _____

SCORE /7 0-3 4-5 6-7

Statistics & Probability

AC9M5ST01

The Bean Experiment

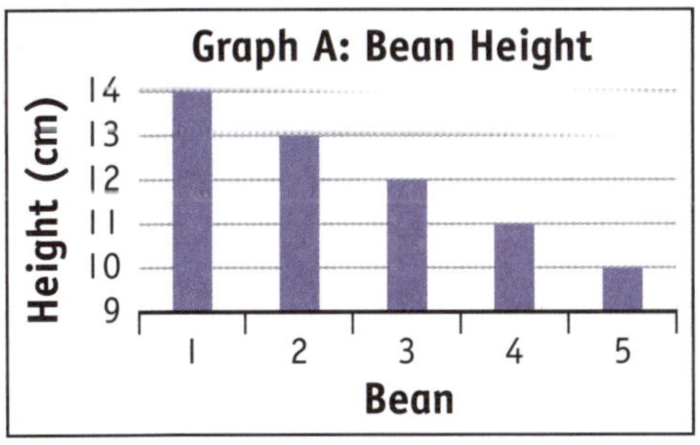

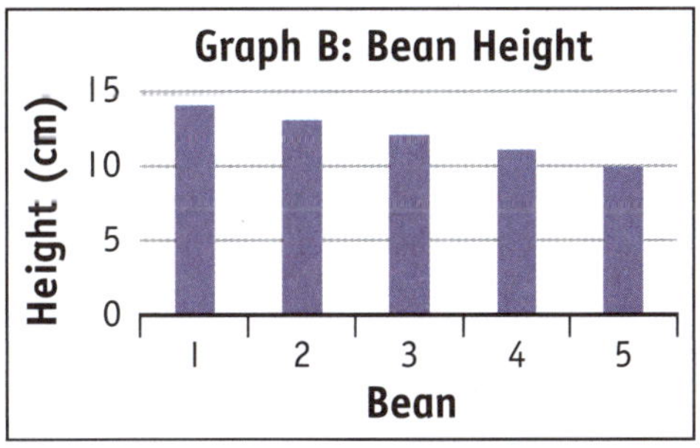

(1) At first glance, which graph shows the greater difference in height between Bean 1 and Bean 5?

◯ Graph A ◯ Graph B

(2) Is the shape of the data the same in Graph A and Graph B?

◯ Same ◯ Different

The graphs show the same data, with the vertical axis showing height in centimetres.

(3) How tall is Bean 1? ______ cm

(4) How tall is Bean 5? ______ cm

(5) If you want to make it look as if the differences in height from Bean 1 to Bean 5 are large, which graph should you use?

◯ Graph A ◯ Graph B

SCORE /5 0-2 3-4 5

AC9M5M04

Drawing Angles

Using the examples to get started, draw the angle specified. Then use your protractor to check your drawing and comment on how close it is to the given angle.

Examples

45°
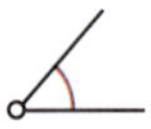
Acute angle

90°

Right angle

135°
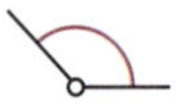
Obtuse angle

180°
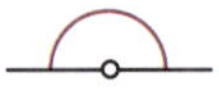
Straight line angle

225°
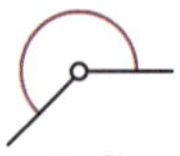
Reflex angle

1. Draw a 30° angle.

   Check and comment:

3. Draw a 270° angle.

   Check and comment:

5. Draw a 80° angle.

   Check and comment:

2. Draw a 120° angle.

   Check and comment:

4. Draw a 90° angle.

   Check and comment:

6. Draw a 200° angle.

   Check and comment:

SCORE /6 0-2 3-4 5-6

NAPLAN*-style Questions

1. The greengrocer has 8 boxes of apples with 48 apples in each box.
   Which of these number sentences should he use to work out how many apples he has?

   ◯ 40 × 8 ◯ 48 × 2 × 2 ◯ 48 + 8 ◯ 48 × 2 × 2 × 2

2. The beginners skateboard ramp needs to be at 25°.
   Which of these ramps is right for beginners?

   ◯ 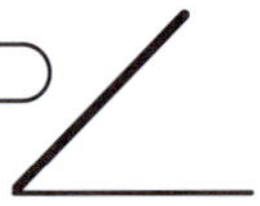 ◯ 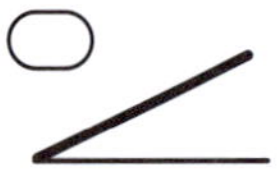 ◯  ◯ 

3. 54 grey cars, 36 blue cars and 18 white cars passed the school gates in one half hour.
   What intervals on the scale would be best to use to show this data on a column graph?

   ◯ 10 ◯ 9 ◯ 3 ◯ 1

SCORE /3 0-1 2 3

TARGETING MATHS HOMEWORK: YEAR 5 © PASCAL PRESS ISBN: 9781925726589

Problem of the Week

AC9M5M04

A Symmetrical Design

I would like you to make a symmetrical design on a grid using only straight lines that go:

Then use arrows to show the direction of travel of the path that goes all the way around your design.

Mark the interior angle at each turn of the path.

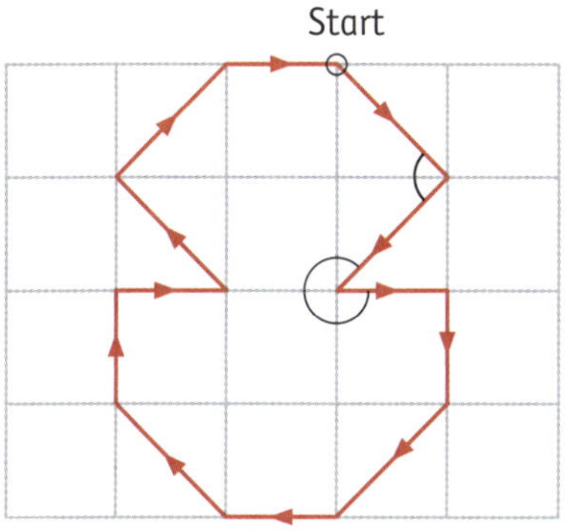

I made this design, but I haven't marked in all the angles of turn, just the first two. Nor have I found the total of the turns.

The Problem

(1) Mark the interior angles of turn in Leo's design.

(2) The total of the angles is ________°.

The Follow-up

(3) Create a symmetrical design of your own using only the types of lines that Mrs Green listed.

(4) Mark in the turn angles and find their total.

(5) Total of the turn angles = ________

Number & Algebra

TERM 2

Fractions

To work out what fraction of a shape has been shaded, it can be helpful to imagine or draw the grid that the shaded section is part of. Here is an example for $\frac{1}{6}$:

$\frac{1}{6}$ One of 6 parts

Draw the grid that the blue section is part of. Write what fraction of the shape is blue.

1. ___ is shaded.

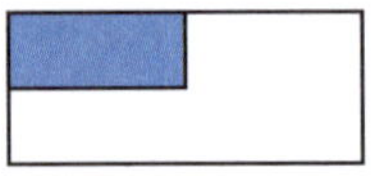

2. ___ is shaded.

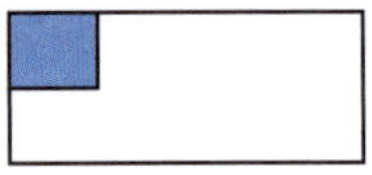

3. ___ is shaded.

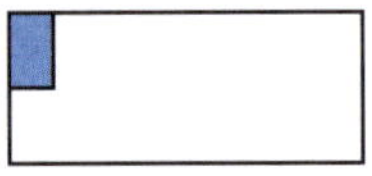

4. ___ is shaded.

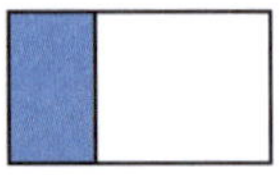

5. ___ is shaded.

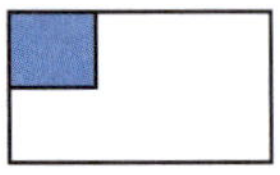

6. ___ is shaded.

7. ___ is shaded.

8. ___ is shaded.

9. ___ is shaded.

Shade each fraction and write an equivalent fraction.

10. 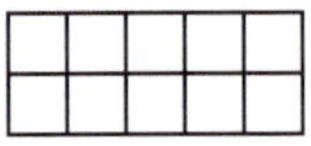

$\frac{2}{10}$ or ___

11. 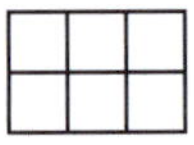

$\frac{4}{6}$ or ___

12. 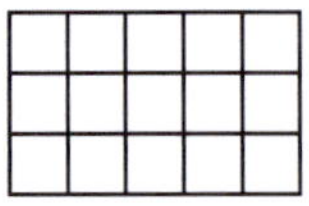

$\frac{3}{15}$ or ___

13. 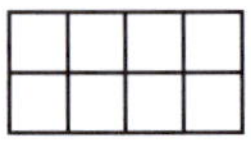

$\frac{2}{8}$ or ___

14. 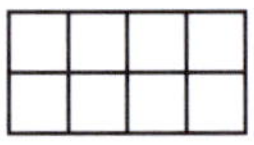

$\frac{6}{8}$ or ___

15. 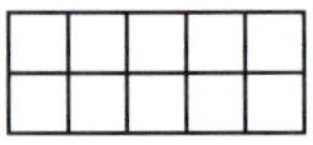

$\frac{5}{10}$ or ___

SCORE /15 0-7 8-13 14-15

Quick Check

What fraction is unshaded?

1. If $\frac{3}{4}$ is shaded, ___ is unshaded.
2. If $\frac{2}{5}$ is shaded, ___ is unshaded.
3. If $\frac{3}{8}$ is shaded, ___ is unshaded.
4. If $\frac{5}{12}$ is shaded, ___ is unshaded.
5. If $\frac{3}{10}$ is shaded, ___ is unshaded.

What is an equivalent fraction?

6. $\frac{4}{8}$ is equivalent to ___
7. $\frac{4}{6}$ is equivalent to ___
8. $\frac{5}{15}$ is equivalent to ___
9. $\frac{6}{12}$ is equivalent to ___
10. $\frac{5}{5}$ is equivalent to ___

SCORE /10 0-4 5-8 9-10

TARGETING MATHS HOMEWORK: YEAR 5 © PASCAL PRESS ISBN: 9781925726589

Number & Algebra

AC9M5N04

Percentages

These are some links between decimals and percentages:

- 0·01 or 1% is 1 in every hundred.
- 0·1 or 10% is 10 in every hundred.
- 0·5 or 50% is 50 in every hundred and is equivalent to a half.
- 0·25 or 25% is 25 in every hundred and is equivalent to a quarter.

What fraction of these shapes is unshaded? Write your answer as a percentage.

1. Unshaded = ____%

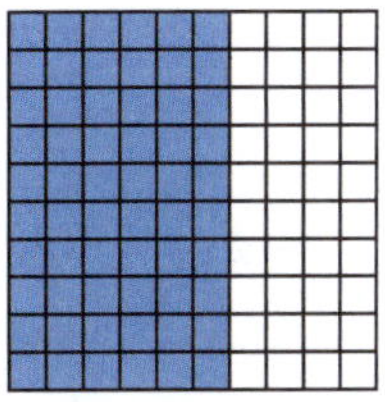

2. Unshaded = ____%

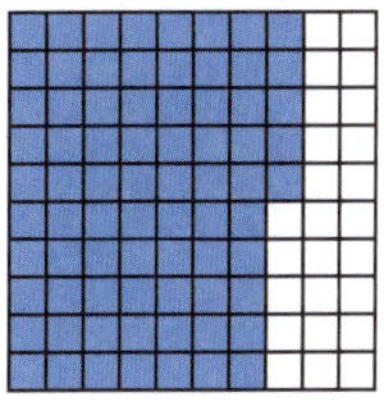

3. Unshaded = ____%

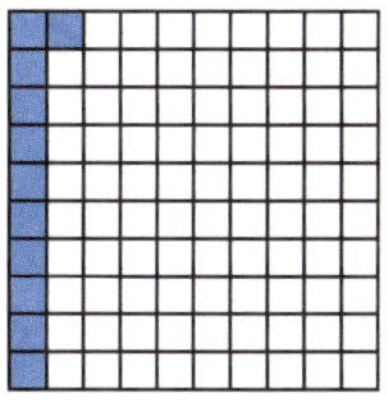

4. Unshaded = ____%

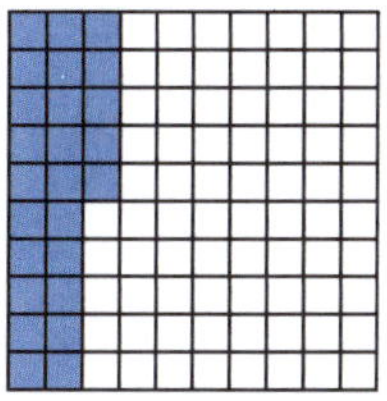

5. Unshaded = ____%

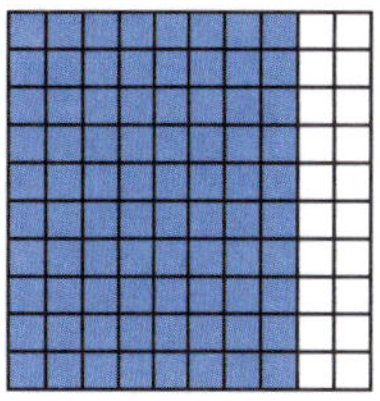

6. Unshaded = ____%

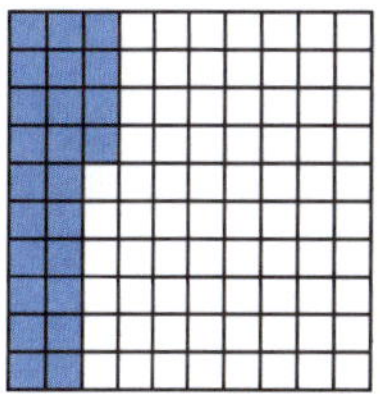

SCORE /6 0-2 3-4 5-6

Statistics & Probability

AC9M5ST02

Graphing Statistics

This table shows the number of people who caught the zoo tram on Wednesday and Thursday.

	9:00	10:00	11:00	Noon	1:00	2:00
Wed.	10	15	20	20	25	10
Thu.	15	15	25	20	15	15

Show this data as a line graph.
Use coloured lines to show the data for each day.

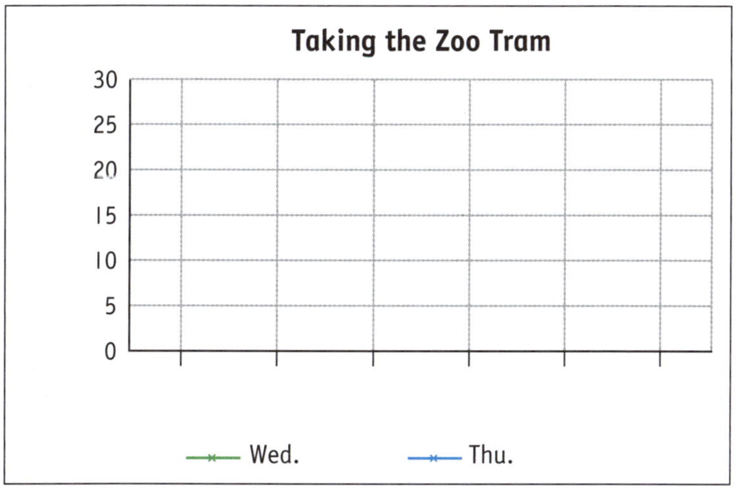

SCORE /6 0-2 3-4 5-6

TARGETING MATHS HOMEWORK: YEAR 5 © PASCAL PRESS ISBN: 9781925726589

AC9M5SP01

TERM 2

Nets in the Real World

Connect each net with its name and real-world object.

① 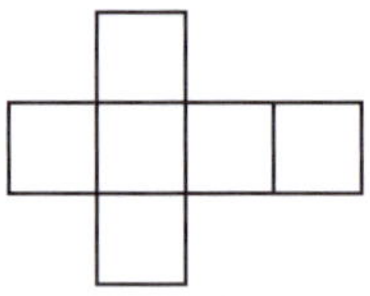 cylinder

② 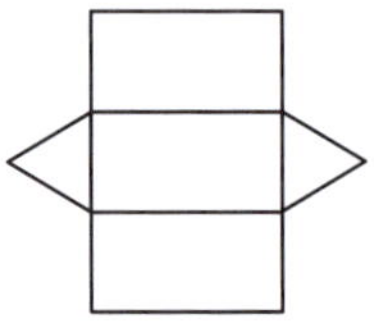 triangular prism

③ 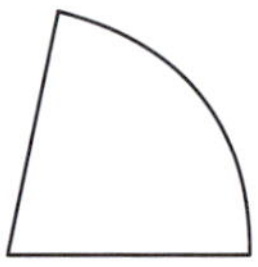 rectangular prism

④ 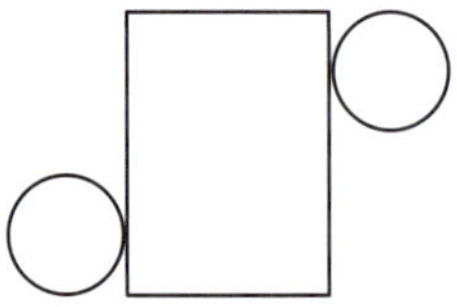 cube

⑤ 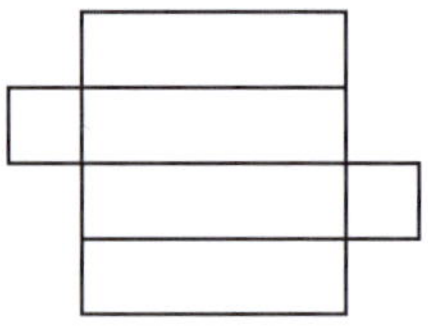 cone

SCORE /5 0-2 3-4 5

NAPLAN*-style Questions

① To make a net for this prism, how many triangles and how many rectangles do you need?

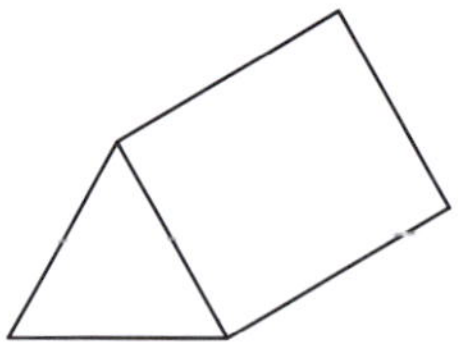

___ triangles ___ rectangles

② This table shows how many students chose each activity at camp.

What percentange of students chose swimming?

Table tennis	Swimming	Nature walk
30	45	25

_____ %

③ 

These cookies each have three choc drops on them.

How many choc drops are on a quarter of the cookies?

____

SCORE /3 0-1 2 3

TARGETING MATHS HOMEWORK: YEAR 5 © PASCAL PRESS ISBN: 9781925726589

Problem of the Week

AC9M5N08

Fill the Board

We have invented a **Fill the Board** percentage game.

We made this spinner. We take turns to spin it and then shade that percentage of our game board.

You have to fill your board exactly – if your spin takes you over 100%, then you can't use that spin.

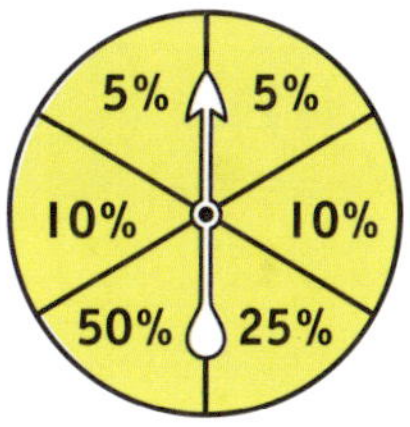

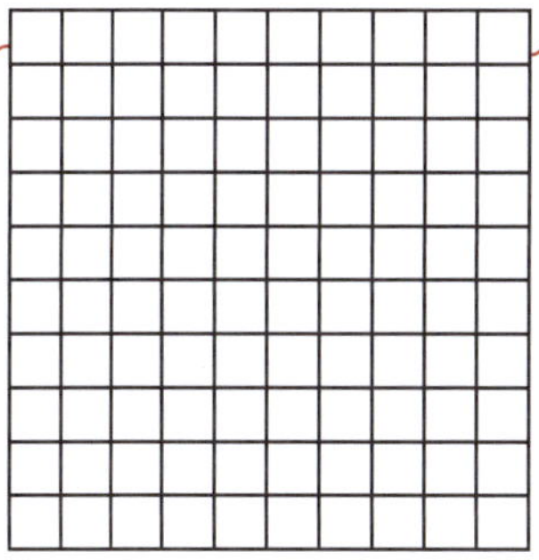

We have just finished seven rounds and we made up some clues about our scores so far.

The Clues

Miro:
This is my board:

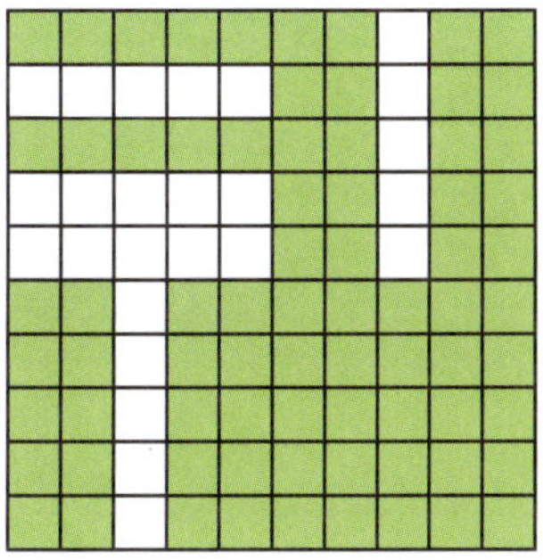

Ella:
My spins were 50%, 5%, 5%, 10%, 5%, 5% and 5%.

Asha:
My spins let me cover $\frac{1}{4}$ of the board twice, $\frac{1}{10}$ of the board once and the rest were 5% spins.

Sam:
I filled half my board on the first spin. The rest of my spins were all the same and I could use all of them.

The Questions

① Who has filled the most of their board at the end of the first seven rounds?

____

② What spins does each player need to fill their board exactly using the smallest number of spins?

Miro: ________

Ella: ________

Asha: ________

Sam: ________

③ Who has the best chance of winning? __________

④ Why do you think that?

____

AC9M5N02, AC9M5N05

Addition of Fractions

Suppose you have to find $\frac{1}{2} + \frac{1}{4}$.
You could use a pie diagram like this:

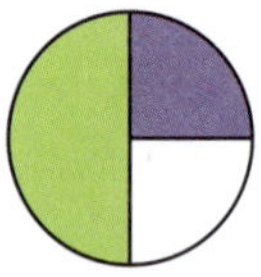

It shows that one-half is **equivalent** to two-quarters and that the total is three-quarters.

Or you could use a number line:

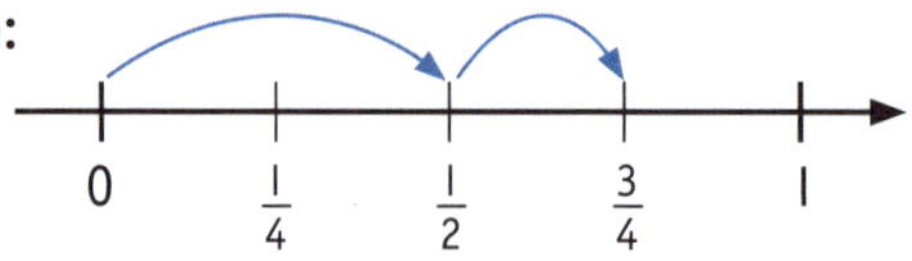

It also shows that one-half is equivalent to two-quarters.

Use equivalence of fractions to solve these additions.
Show your thinking on the pie diagram and on the fraction line.

Addition	Pie diagram	Fraction line	Answer
$\frac{1}{3} + \frac{1}{6}$	①	② 0 … 1	
$\frac{1}{5} + \frac{3}{10}$	③	④ 0 … 1	
$\frac{3}{8} + \frac{1}{4}$	⑤	⑥ 0 … 1	
$\frac{1}{2} + \frac{3}{10}$	⑦	⑧ 0 … 1	
$\frac{3}{4} + \frac{3}{8}$	⑨	⑩ 0 … 1	

SCORE /10 0-4 5-8 9-10

Quick Check

① $\frac{1}{4} + \frac{1}{4} = \frac{2}{4}$ or $\frac{1}{2}$

② $\frac{2}{5} + \frac{3}{5} =$ — or ___

③ $\frac{1}{4} + \frac{3}{4} =$ — or ___

④ $\frac{3}{8} + \frac{3}{8} =$ — or —

⑤ $\frac{3}{10} + \frac{2}{10} =$ — or —

⑥ $\frac{3}{5} + \frac{4}{10} =$ — or ___

⑦ $\frac{2}{3} + \frac{3}{6} =$ — or ___ —

⑧ $\frac{3}{8} + \frac{6}{8} =$ — or ___ —

⑨ $\frac{3}{10} + \frac{3}{10} =$ — or —

⑩ $\frac{3}{6} + \frac{1}{2} =$ — or ___

SCORE /10 0-4 5-8 9-10

TARGETING MATHS HOMEWORK: YEAR 5 © PASCAL PRESS ISBN: 9781925726589

Subtraction of Fractions

You can also use equivalence of fractions to help with subtraction.

To subtract $\frac{1}{6}$ from $\frac{1}{3}$, convert $\frac{1}{3}$ into $\frac{2}{6}$.

Then do the subtraction: $\frac{2}{6} - \frac{1}{6} = \frac{1}{6}$

Convert and complete these subtractions.

1. $\frac{2}{3} - \frac{1}{6} = \_\_ - \_\_ = \_\_$
2. $\frac{3}{8} - \frac{1}{4} = \_\_ - \_\_ = \_\_$
3. $\frac{1}{2} - \frac{3}{10} = \_\_ - \_\_ = \_\_$
4. $\frac{3}{4} - \frac{3}{8} = \_\_ - \_\_ = \_\_$
5. $\frac{2}{5} - \frac{1}{10} = \_\_ - \_\_ = \_\_$
6. $\frac{5}{8} - \frac{1}{4} = \_\_ - \_\_ = \_\_$
7. $\frac{3}{5} - \frac{3}{10} = \_\_ - \_\_ = \_\_$
8. $\frac{3}{4} - \frac{1}{8} = \_\_ - \_\_ = \_\_$
9. $\frac{3}{3} - \frac{5}{6} = \_\_ - \_\_ = \_\_$
10. $\frac{3}{4} - \frac{5}{12} = \_\_ - \_\_ = \_\_$

SCORE /10 0-4 5-8 9-10

Statistics & Probability

AC9M5P01

Fractions for Chance

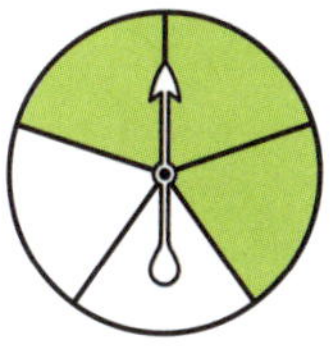

With this spinner, the chance of spinning green is 3 out of 5.
This chance can be shown on a probability line:

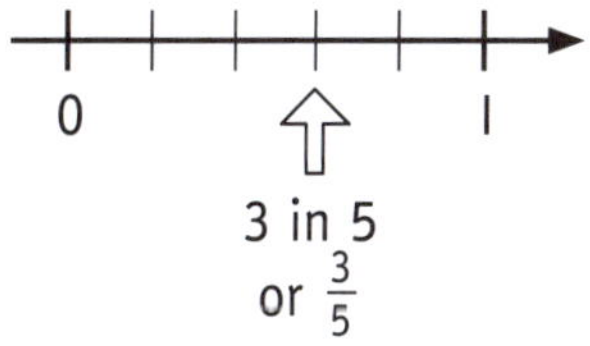

Show the chance of spinning green on the probability lines.

1.
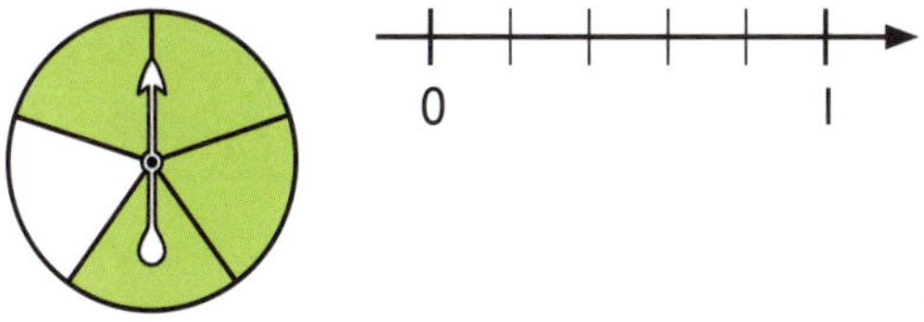

2.
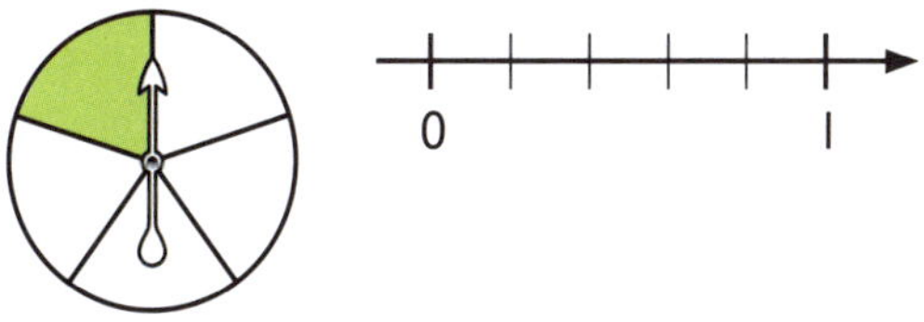

3.
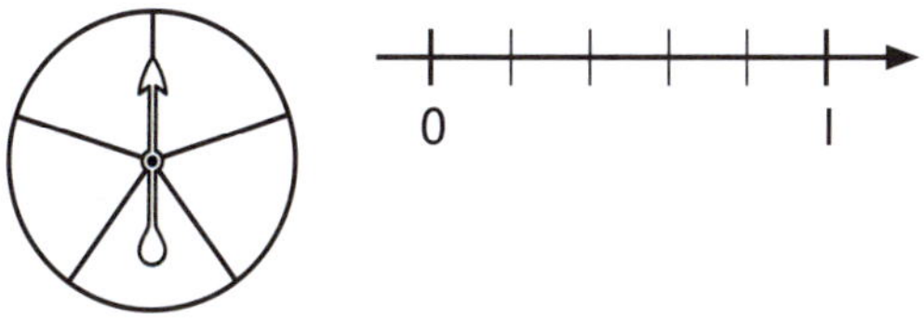

4.
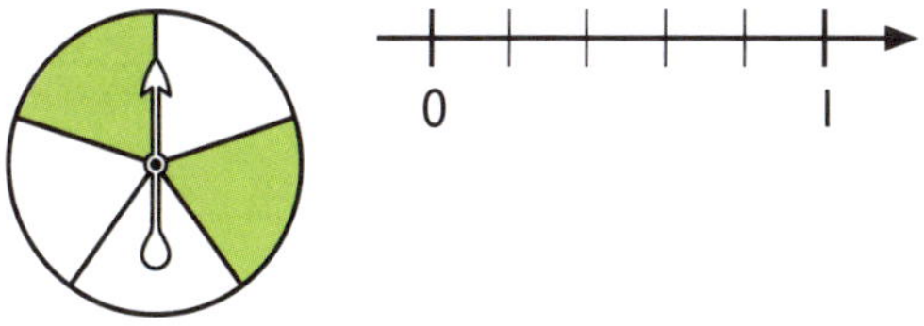

5.
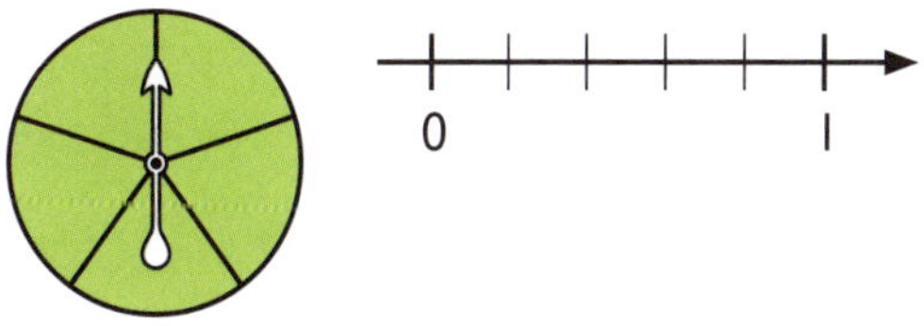

SCORE /5 0-2 3-4 5

TERM 2

Area and Perimeter

Use this shape to answer the following questions.

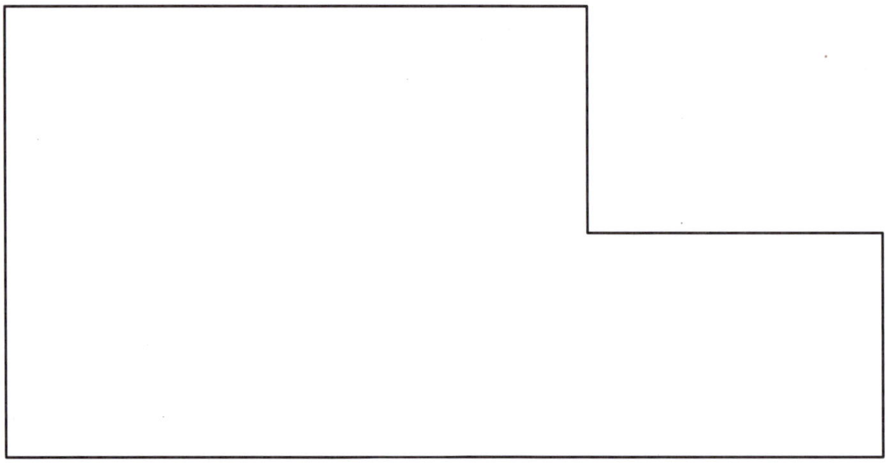

1. Use your ruler to find the perimeter of the shape.

   The perimeter is _______ cm.

2. What is the area of the shape in cm^2? _______ cm^2
3. If the shape were twice the size, do you think the perimeter would be doubled? _____
4. If the shape were twice the size, do you think the area would be doubled? _____

Work out the perimeter and area for a shape twice the size. You can draw the new shape to help.

5. Perimeter = ____ cm
6. Area = ____ cm^2

SCORE /6 0-2 3-4 5-6

NAPLAN*-style Questions

1. Ella had 48 blocks. She built a tower with a quarter of her blocks.

   How many blocks did Ella use to build her tower?

   [   ] blocks

2. Which of these shapes has the longest perimeter?

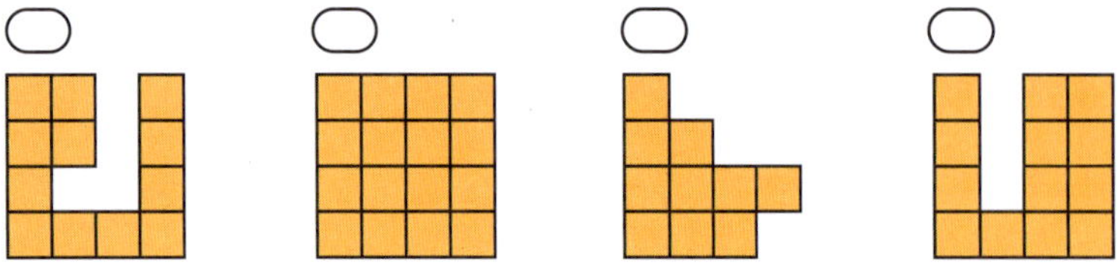

3. Which of these spinners has a 1 in 6 chance of spinning green?

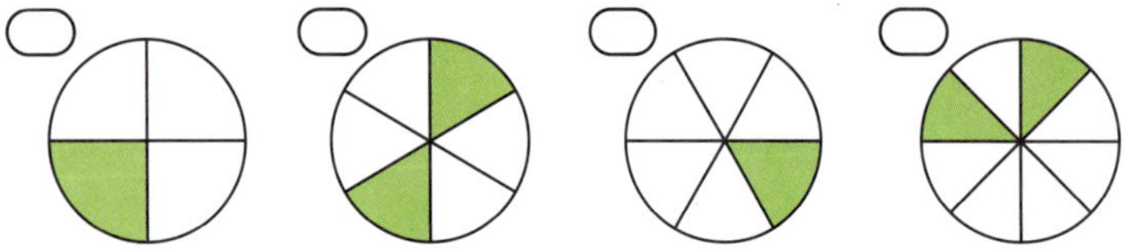

SCORE /3 0-1 2 3

TARGETING MATHS HOMEWORK: YEAR 5 © PASCAL PRESS ISBN: 9781925726589

Problem of the Week

UNIT 12

AC9M5N02, AC9M5N05

The Pizza Party

Jade was late arriving at the pizza party and she found that there was not much pizza left. Four pizzas had been ordered, but her friends had eaten well, and poor Jade was not happy.

Who ate how much pizza?

- Leo ate $\frac{3}{8}$ of a pizza.
- Asha ate twice as much as Leo.
- Miro ate half a pizza more than Leo and a quarter more than Ella.
- Sam was really hungry and ate twice as much as Leo.
- Ravi wasn't very hungry and only ate half as much as Asha.

What was left for Jade?

1. Use these pizza shapes to show how much each person ate.
2. How much was left for Jade? ____

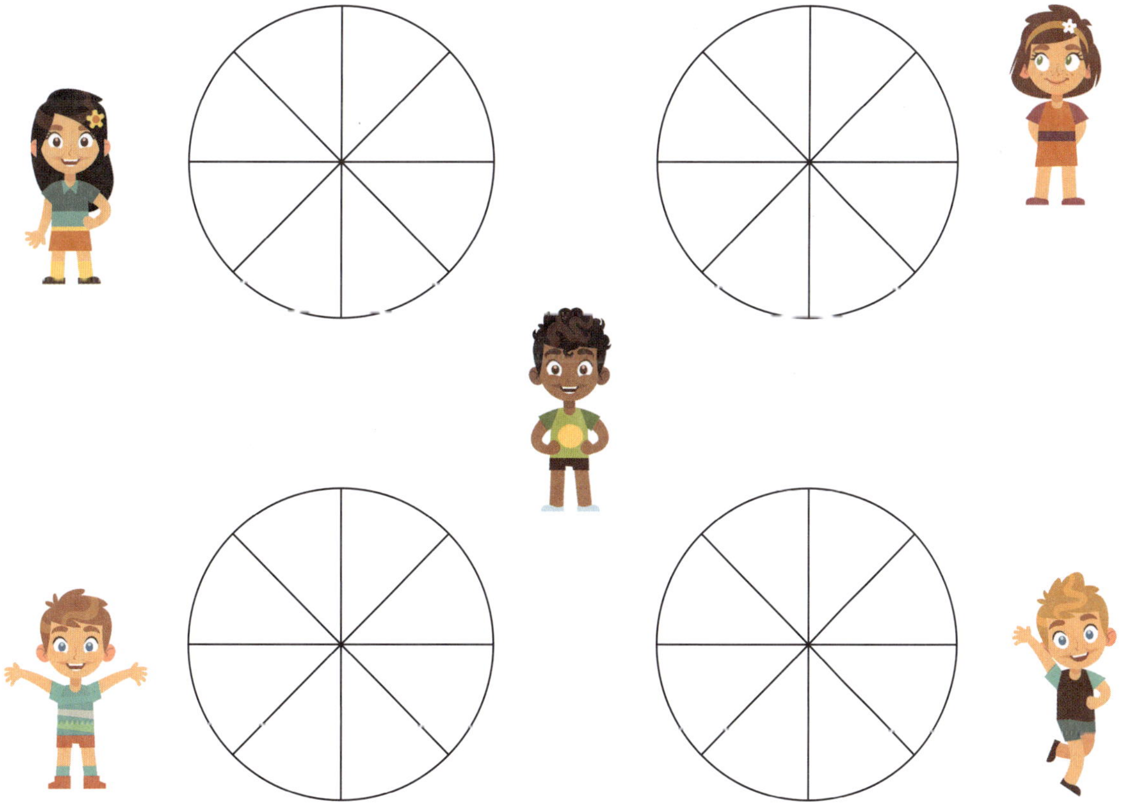

TERM 2

AC9M5N07

Division with Remainders

When we work out the division 6 ÷ 2, the answer is the whole number 3.
We say that 2 divides 6 evenly as there is nothing left over (no remainder).
But for 13 ÷ 2, the answer is 6 with 1 left over, so we write: **13 ÷ 2 = 6 r 1**.
We say that as 13 divided by 2 equals 6 remainder 1.

Carry out these divisions and write the remainder each time.

1. 16 ÷ 3 = ___ r ___
2. 15 ÷ 4 = ___ r ___
3. 26 ÷ 4 = ___ r ___
4. 19 ÷ 3 = ___ r ___

Sometimes we treat the remainder differently.
Read this question:
How many bags of 6 marbles can be made from 13 marbles?
Here, we don't name the remainder because the answer we need is **six bags**.

Solve these word problems.

5. 15 oranges were shared between 4 nets.
   How many oranges were left over?
   ___ ÷ ___ = ___ r ___
   ___ oranges were left over.
6. How many nets are needed to put 5 onions in each net when there are 37 onions available?
   ___ ÷ ___ = ___ r ___
   ___ nets will be needed.
7. The box of 26 oranges was made into 5 equal-sized nets.
   How many oranges were left over?
   ___ ÷ ___ = ___ r ___
   ___ oranges were left over.
8. How many nets of 6 apples can be made from 27 apples?
   ___ ÷ ___ = ___ r ___
   ___ nets can be made.
9. The shopkeeper made 8 nets of 6 avocadoes and had 2 avocadoes left over.
   How many avocadoes did they start with?
   ___ ÷ ___ = ___ r ___
   There were ___ avocadoes altogether.

SCORE /9 0-4 5-7 8-9

Quick Check

Complete these division questions and write the remainders.

1. 37 ÷ 5 = ___ r ___
2. 42 ÷ 9 = ___ r ___
3. 64 ÷ 7 = ___ r ___
4. 38 ÷ 4 = ___ r ___
5. 57 ÷ 8 = ___ r ___
6. 49 ÷ 6 = ___ r ___
7. 73 ÷ 6 = ___ r ___
8. 93 ÷ 9 = ___ r ___
9. 79 ÷ 11 = ___ r ___
10. 59 ÷ 6 = ___ r ___

SCORE /10 0-4 5-8 9-10

TARGETING MATHS HOMEWORK: YEAR 5 © PASCAL PRESS ISBN: 9781925726589

Number & Algebra

Division with Fraction Remainders

Sometimes the answer to a division includes a fraction. For example:

If three mini-pizzas are to be divided between two people, how much will each person get?

The answer is $1\frac{1}{2}$ mini-pizzas each.

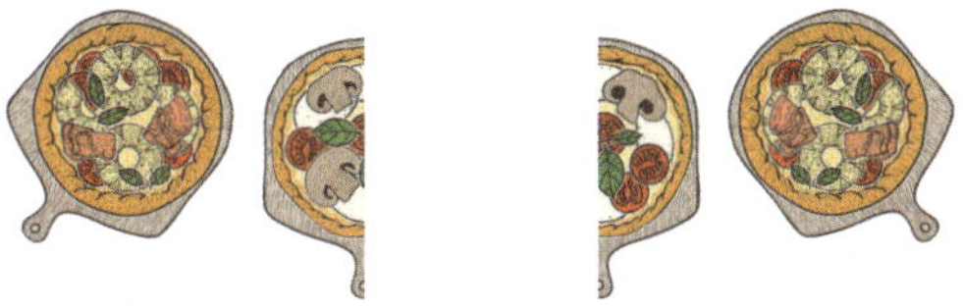

Answer these division questions and include the fraction part each time.

1. 5 kilograms of potatoes are to be shared between 2 bags.

   How many kilograms of potatoes will be in each bag?

   ___ ÷ ___ = _____ kg

2. 5 metres of elastic are to be cut into 4 strips.

   How long will each strip be?

   ___ ÷ ___ = _____ m

3. 4 litres of juice are to be poured equally into 3 containers.

   How much juice will be in each container?

   ___ ÷ ___ = _____ L

4. 9 metres of fencing are to be made into 6 identical fence panels.

   How long will each fence panel be?

   ___ ÷ ___ = _____ m

5. 7 bottles of juice are to be shared between 4 tall glasses.

   How much juice will be in each glass?

   ___ ÷ ___ = _____ bottles of juice

SCORE /5 0-2 3-4 5

Statistics & Probability

AC9M5ST03

Sport on TV

100 people were surveyed about which sport they preferred to watch on television.

- 35% said football.
- 25% said soccer.
- 10% said golf.
- The rest said cricket.

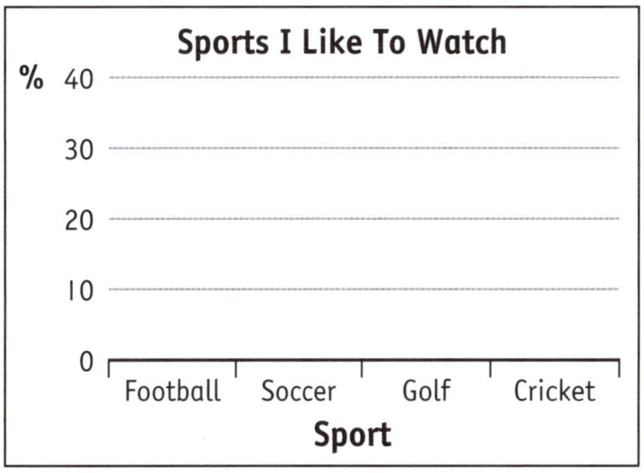

1. Complete the graph to match the survey data.
2. How many people preferred football? _____
3. How many people preferred soccer? _____
4. How many people preferred golf? _____
5. How many people preferred cricket? _____

SCORE /5 0-2 3-4 5

Capacity

Chef designed a new set of measuring cups where one whole cup holds **360 mL**.

She finds it makes the cups easy to work with.

Trainee chefs have to answer the following questions based on Chef's new set of cups.

Answer the questions to see if you have what it takes to be a trainee chef.

1. How much does a quarter cup hold? _____ mL
2. How much does a third cup hold? ______ mL

Which combination of cups would you use to measure these capacities?

3. 210 mL ________________
4. 270 mL ________________
5. 420 mL ________________

SCORE /5 0-2 3-4 5

NAPLAN*-style Questions

1. There were 15 strawberries in a punnet. Will shared them equally with a friend. How many strawberries were left over?

   ☐ strawberries

2. A juice carton holds 600 mL of juice. Ravi filled three 180 mL glasses. How much juice was left in the container?

   ◯ 240 mL ◯ 180 mL ◯ 120 mL ◯ 60 mL

3. Sam collected data about her screen time last week. She made this graph of how much time she spent on each activity.

   25% of the time was spent playing games.

   What percentage of Sam's screen time was spent on homework?

   ☐ %

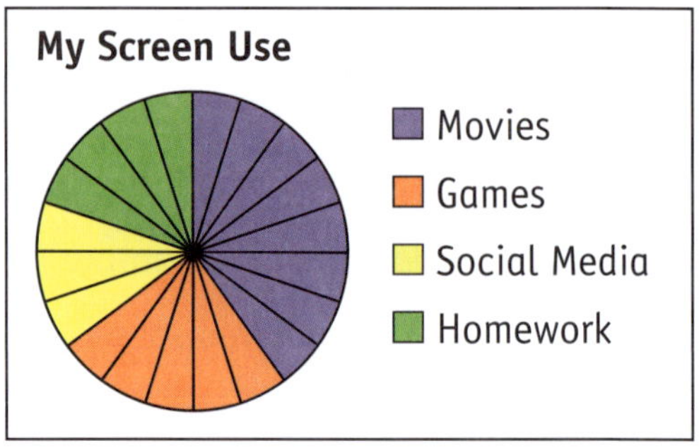

SCORE /3 0-1 2 3

TARGETING MATHS HOMEWORK: YEAR 5 © PASCAL PRESS ISBN: 9781925726589

Calculator Magic

Here is a calculator sequence for you to try.

(1) Think of a 2-digit number (between 10 and 99) and write it down.

10s	1s

(2) Press MC to clear the memory.

(3) Enter your number in the calculator and press M+.

(4) Add the reverse of your number, then divide by 11.

(5) Multiply by 8, add the original number and divide by 9.

(6) Multiply by 7, add the original number and divide by 8.

(7) Multiply by 6, add the original number and divide by 7.

(8) Multiply by 5, add the original number and divide by 6.

(9) Multiply by 4, add the original number and divide by 5.

(10) Multiply by 3, add the original number and divide by 4.

(11) Multiply by 2, add the original number and divide by 3.

(12) Multiply by 1, add the original number and divide by 2.

(13) Add the first digit of your original number (that's the 10s digit).

(14) Write the number in the display in the first box below.

(15) Repeat the above steps four more times and record the last number in the display each time.

(16) What can you say about the number that is left in your display?

____

I have a suggestion that might help!
Try this key sequence to make sure you don't make any mistakes.

×	8	+	MR	÷	9	=
×	7	+	MR	÷	8	=
×	6	+	MR	÷	7	=

... and so on.

TERM 2

Round and Adjust

When trying to work out the cost of 3 apples at $1·85 each, we can round the amount and then adjust to find the answer.

$1·85 × 3
↓ Rounding by adding 15c
$2·00 × 3 = $6·00
↓ Adjust by subtracting 15c × 3 = 45c
$5·55

Complete these round and adjust multiplications.

1. $8·75 × 6
↓ Rounding by adding ____
$______ × __ = _______
↓ Adjust by subtracting ___ × __ = ____
$______

2. $10·80 × 5
↓ Rounding by adding ____
$______ × __ = _______
↓ Adjust by subtracting ___ × __ = ____
$______

3. $3·90 × 7
↓ Rounding by adding ____
$______ × __ = _______
↓ Adjust by subtracting ___ × __ = ____
$______

4. $11·80 × 8
↓ Rounding by adding ____
$______ × __ = _______
↓ Adjust by subtracting ___ × __ = ____
$______

You need to round down for the next two examples.

5. $9·05 × 7
↓ Rounding by subtracting ____
$______ × __ = _______
↓ Adjust by adding ___ × __ = ____
$______

6. $12·30 × 4
↓ Rounding by subtracting ____
$______ × __ = _______
↓ Adjust by adding ___ × __ = ____
$______

SCORE /6 0-2 3-4 5-6

Quick Check

Answer these using round and adjust in your head.

1. 6 at $4·90 each _______
2. 3 at $3·80 each _______
3. 7 at $2·95 each _______
4. 6 at $4·75 each _______
5. 8 at $4·80 each _______
6. 3 at $6·15 each _______
7. 4 at $3·25 each _______
8. 6 at $1·20 each _______
9. 5 at $4·15 each _______
10. 8 at $3·50 each _______

SCORE /10 0-4 5-8 9-10

TARGETING MATHS HOMEWORK: YEAR 5 © PASCAL PRESS ISBN: 9781925726589

Number & Algebra

AC9M5N09

Practical Word Problems

For each of these problems, write the matching number sentences and the answer.

1. Sam bought 4 apples at 95 cents each and paid with a $5 note.
   How much change did she get?
   ______ × ______ = ______
   ______ – ______ = ______

2. Leo is winning the 'Race to $100' game. He has won three lots of $15 and a $20 and a $15 token.
   How far from $100 is he?
   ______ × ______ + ______ = ______
   ______ – ______ = ______

3. Ella shared $24 fairly between herself and three friends.
   How much money did Ella give away?
   ______ ÷ ______ = ______
   ______ × ______ = ______

4. Jade had a punnet of 24 strawberries. She ate half of them and then later she ate half of what was left. Then she gave half of the remaining strawberries to her brother.
   How many strawberries did Jade still have?
   ______ × ______ = ______
   ______ × ______ = ______

5. Miro had $12·50 left after spending $3·50 a day on his school lunches every day this week.
   How much money did Miro have to begin with?
   ______ × ______ = ______
   ______ + ______ = ______

SCORE /5 0-2 3-4 5

Statistics & Probability

AC9M5ST01

Finding the Mode

Will and Ravi have been rolling a dice, recording the results of their throws and finding the mode. Will rolled 6, 4, 2, 5, 2, 4, 6, 3, 2.
To find the mode, he sorted the numbers and set them out in order like this:
2, 2, 2, 3, 4, 4, 5, 6, 6.
Then he counted how many of each number he had thrown.
He had thrown more 2s than any other number, so 2 was his mode.

Reorder these dice strings and find the modes.

1. 4, 3, 2, 4, 1, 4, 3, 5, 6
   ______________________ The mode is ____.
2. 1, 6, 3, 5, 4, 4, 5, 5, 3
   ______________________ The mode is ____.
3. 5, 5, 6, 1, 2, 2, 3, 2, 2
   ______________________ The mode is ____.
4. 3, 6, 5, 4, 3, 2, 1, 3, 5
   ______________________ The mode is ____.
5. 1, 2, 5, 6, 1, 3, 3, 5, 3
   ______________________ The mode is ____.

SCORE /5 0-2 3-4 5

Calling Times

I like to video chat with my cousins in London.
But when I do, I need to check what time it is on the other side of the world.
In summer, London is $10\frac{1}{2}$ hours behind the time in Adelaide, where I live.

Use these timelines to help you answer the questions below.

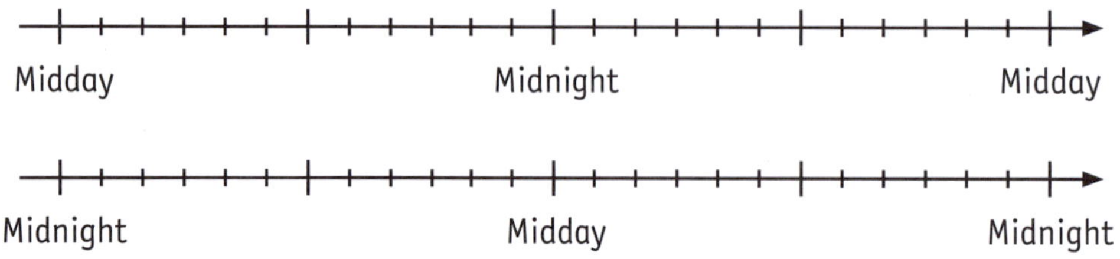

What is the time in London at these Adelaide times? Use 24-hour time format.

1. Adelaide time: 01:30 London time: ______
2. Adelaide time: 15:40 London time: ______
3. Adelaide time: 07:50 London time: ______
4. Adelaide time: 23:30 London time: ______

What is the time in Adelaide at these London times?

5. London time: 03:30 Adelaide time: ______
6. London time: 16:30 Adelaide time: ______
7. London time: 21:30 Adelaide time: ______
8. Which of the above times would be best for me to video chat with my cousins?

____

SCORE /8 0-3 4-6 7-8

NAPLAN*-style Questions

1. A newsagent sells packets of collector cards for $2·75.
   How much change will Sam get from $10 after buying 3 packets of cards?
   - ◯ $8·25 ◯ 25c ◯ $2·75 ◯ $1·75
2. Jade went to the swimming pool at 11:00 am.
   She left the pool at 2:38 pm.
   How long was Jade at the pool?
   - ◯ 2 h 38 min ◯ 2 h 48 min ◯ 3 h 48 min ◯ 3 h 38 min
3. 3 metres of ribbon was cut into quarters to make bows.
   How long was each cut piece of ribbon?
   - [ ] cm

SCORE /3 0-1 2 3

TARGETING MATHS HOMEWORK: YEAR 5 © PASCAL PRESS ISBN: 9781925726589

Problem of the Week

AC9M5M03

Reading the Paper

My grandma spent 2 hours reading the paper last week, but no more than half an hour each day.

Problem:
How long did my grandma spend reading the paper each day?

Clues

1. On Tuesday and Saturday, she read for the same length of time.
2. On Wednesday, she spent three times as long as on Monday, and she did the same on Sunday.
3. On Thursday, she spent the same as Monday and Tuesday put together.
4. On Friday, she spent five minutes less than on Wednesday.
5. She spent twice as much time reading the paper on Monday as she did on Tuesday.

Workspace

The Solution

Day	Time (minutes)
Monday	
Tuesday	
Wednesday	
Thursday	
Friday	
Saturday	
Sunday	
Total	

AC9M5A01, AC9M5A02

TERM 2

Multiplication and Division with Unknowns

Mrs Green likes to set us questions that have unknown numbers in them.

I do enjoy working out what these unknown numbers might be and would like to share my method of working with you.

Here is an example:

$6 \times \square = \square \div 2 = 18$

I know that each of the first two parts in the equation have to equal 18.

So I start by asking, "What multiplied by 6 gives me 18?"

And then, "What divided by 2 gives me 18?"

I'm sure you've got the answer by now!

$6 \times 3 = 36 \div 2 = 18$

Complete these equations by finding the unknowns.

1. $4 \times \square = \square \div 3 = 12$
2. $\square \times 7 = 56 \div \square = 28$
3. $54 \div \square = \square \times 6 = 18$
4. $\square \times 8 = 64 \div \square = 32$
5. $110 \div \square = 11 \times \square = 11$
6. $72 \div \square = \square \div 8 = 6$
7. $\square \times 5 = 130 \div \square = 65$
8. $3 \times \square = \square \div 3 = 60$
9. $9 \times \square = 108 \div \square = 54$
10. $18 \times \square = 108 \div \square = 36$
11. $99 \div \square = 88 \div \square = 11$
12. $88 \div \square = \square \div 3 = 22$

SCORE /12 0-5 6-10 11-12

Quick Check

Complete these equations using related multiplication facts.

But Mrs Green always says that your unknown numbers must never be 1.
For example, the answer to the first question is not allowed to be $3 \times 4 = 12 \div 1$.

1. $3 \times \square = 12 \div \square$
2. $\square \times 3 = 24 \div \square$
3. $\square \times 6 = 36 \div \square$
4. $\square \times 5 = 45 \div \square$
5. $\square \times 4 = 32 \div \square$
6. $9 \times \square = 54 \div \square$
7. $7 \times \square = 56 \div \square$
8. $4 \times \square = 56 \div \square$
9. $\square \times 4 = 84 \div \square$
10. $\square \times \square = 36 \div 9$

SCORE /10 0-4 5-8 9-10

TARGETING MATHS HOMEWORK: YEAR 5 © PASCAL PRESS ISBN: 9781925726589

Factors

I'm sure you know that 6 has four factors: 1, 2, 3 and 6.

But every number has 1 and itself as factors and so we use the term **proper factor** to refer to those factors that are not 1 or the number itself.

Yes, 2 and 3 are the proper factors of 6.

Now for the question.

Use the spreadsheet to answer the questions.

1. Add a tick to a cell if the number in Column B is a proper factor of the number in Row 2.
2. Which number between 20 and 30 has the most proper factors? ______
3. Which numbers between 20 and 30 have only two proper factors? ______
4. Which numbers in the range 21–29 are divisible by 7?
____
5. Which number in the range 21–29 is only divisible by 5? ______
6. What are the proper factors of 24?
____
7. Which numbers in the range 21–29 have no proper factors?
____

	A	B	C	D	E	F	G	H	I	J	K
1											
2			21	22	23	24	25	26	27	28	29
3		2									
4		3									
5		4									
6		5									
7		6									
8		7									
9		8									
10		9									
11		10									
12		11									
13		12									
14		13									
15		14									
16		Total									

How to make the spreadsheet (if you want to have a go yourself)

- Enter 2 and 3 in cells B3 and B4.
- Select B3:B4 and use the Fill Handle to fill down to B15.

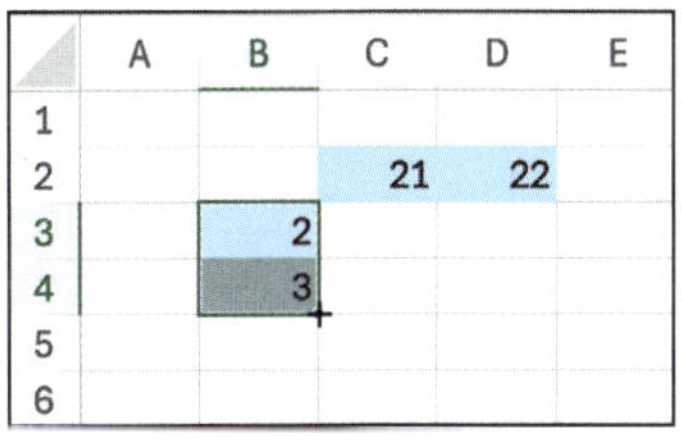

- Enter 21 and 22 in C2 and D2.
- Select C2:D2 and use the Fill Handle to fill across to K2.

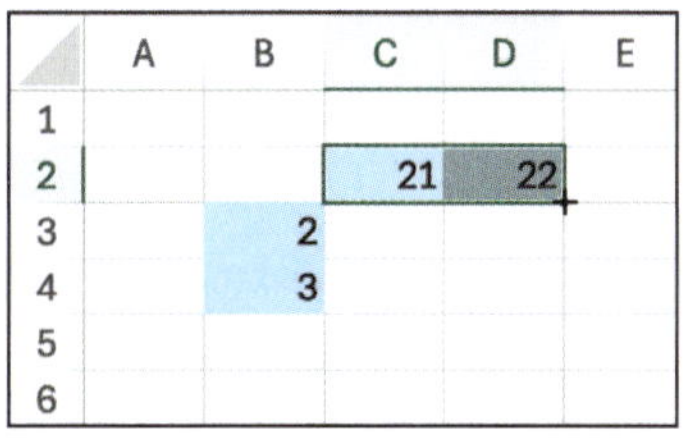

- Enter this formula in cell C3:
  =IF(MOD(C$2, $B3)=0,1,0)
- Use the Fill Handle of C3 to add the formula to the whole table C3:K15.
- This will place a 1 in every cell in which the factors occur.

- To complete the spreadsheet, make B16 = Total.
- Select C16 and enter the formula:
  =SUM(C3:C15)
- Use the Fill Handle to add the formula to cells D16 to K16. This will show the answers to Questions 2, 3 and 7.

SCORE /7 0-3 4-5 6-7

TERM 2

AC9M5SP02

Coordinates

Start at (0, 0) and make a journey to collect each shape.

Write the coordinates of the shape you collect and the moves needed to get to it.

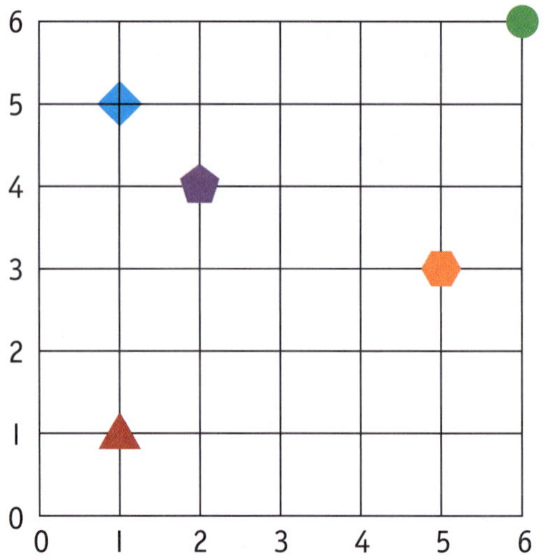

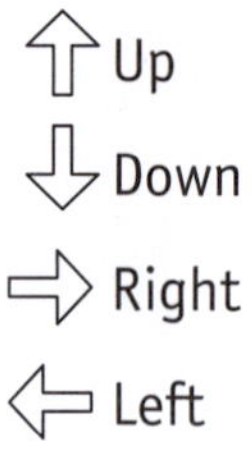

1. To collect the at (__, __) from (0, 0), go right 1 and up ___.
2. To collect the at (__, __) from the at (__, __), go ______________________.
3. To collect the at (__, __) from the at (__, __), go ______________________.
4. To collect the at (__, __) from the at (__, __), go ______________________.
5. To reach the exit at (__, __) from the at (__, __), go ______________________.

SCORE /5 0-2 3-4 5

NAPLAN*-style Questions

1. What is the number that will make this division correct?

   45·6 ÷ ☐ = 4·56

2. Miro has entered the formula **=B2+2** in cell B3 and will use the Fill Handle to copy the formula to cell B9.

   What will be the value in cell B9 when he has done that?

   B9 = ☐

	A	B	C
1			
2		1	
3		3	
4			
5			

3. Which coordinates do not match a corner of the shape in the grid?

   ⬭ (2, 1) ⬭ (4, 2)
   ⬭ (4, 5) ⬭ (3, 3)
   ⬭ (5, 1)

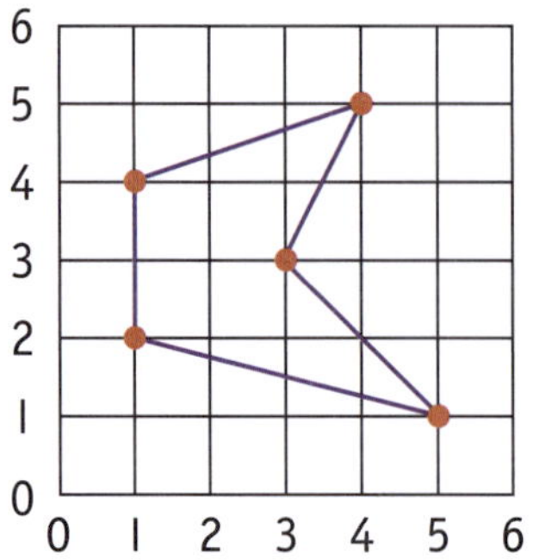

SCORE /3 0-1 2 3

TARGETING MATHS HOMEWORK: YEAR 5 © PASCAL PRESS ISBN: 9781925726589

Problem of the Week

AC9M5SP01

Collect the Treasure and Escape

We have just made up this Collect the Treasure and Escape game.
The game is played on the board below.
The aim is to collect as many treasures as you can.

These are the treasures and their positions on the board:

 **(2, 2)**  **(3, 4)**  **(4, 4)**  **(5, 2)**

At the start, you are on (0, 0). To finish, you must get to (6, 6).
We have also made a set of cards that show the direction that you should move and how far you can go. There are 3 of each card.

3 each of Set 1: ↑ ↓ → ← 3 each of Set 2: 1 2 3 4

- To begin, the cards are shuffled and each player is dealt 6 cards from Set 1 and 6 cards from Set 2.
- The player with the higher total of Set 2 cards has the first turn.
- After that, players take turns to use one of each type of card to move around the board.
- If possible, they land on a treasure that they collect and remove from the game board.

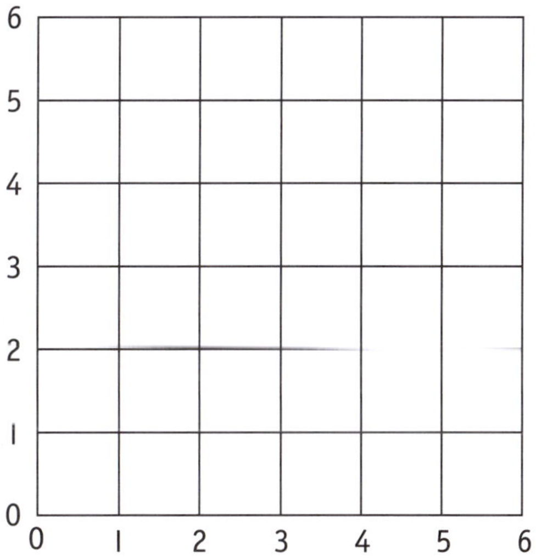

The Challenge

① Place the treasures on the board.

② Choose a set of 6 cards of each type (direction and distance) and record them here.

and

③ Can you move from (0, 0) to (6, 6) and collect all the treasures with the cards that you chose?

④ Even if your answer to Question 3 was yes, choose a **different** set of cards that will enable you to collect all the treasures. Record your new cards here.

and

TARGETING MATHS HOMEWORK: YEAR 5 © PASCAL PRESS ISBN: 9781925726589

TERM 2

Decimals as Fractions

Decimals are fractions in a different form.
But you can change one type of fraction into the other quite easily.
Here are some examples:

0·35 can be changed to $\frac{35}{100}$ or to $\frac{3}{10} + \frac{5}{100}$.

0·335 can be changed to $\frac{335}{1000}$ or to $\frac{3}{10} + \frac{3}{100} + \frac{5}{1000}$.

0·305 can be changed to $\frac{305}{1000}$ or to $\frac{3}{10} + \frac{5}{1000}$.

Show how to change each of these decimals into fractions.

1. 0·27 = ______ or ____ + ____
2. 0·76 = ______ or ____ + ____
3. 0·138 = ______ or ____ + ____ + ____
4. 0·348 = ______ or ____ + ____ + ____
5. 0·257 = ______ or ____ + ____ + ____
6. 0·158 = ______ or ____ + ____ + ____
7. 0·708 = ______ or ____ + ____ + ____
8. 0·109 = ______ or ____ + ____ + ____
9. 0·018 = ______ or ____ + ____ + ____
10. 0·507 = ______ or ____ + ____ + ____
11. 0·035 = ______ or ____ + ____ + ____
12. 0·123 = ______ or ____ + ____ + ____

Show how to write these fractions as decimals.

13. $\frac{1}{10} + \frac{2}{100}$ = ______
14. $\frac{3}{10} + \frac{4}{1000}$ = ______
15. $\frac{5}{100} + \frac{6}{1000}$ = ______
16. $\frac{12}{10} + \frac{5}{100}$ = ______
17. $\frac{35}{10} + \frac{2}{100}$ = ______
18. $\frac{27}{10} + \frac{27}{1000}$ = ______

SCORE /18 0-8 9-16 17-18

Quick Check

Write these combinations of fractions as decimals.

1. $\frac{4}{10} + \frac{3}{100} + \frac{5}{1000}$ = ______
2. $\frac{7}{10} + \frac{5}{1000}$ = ______
3. $\frac{8}{100} + \frac{6}{1000}$ = ______
4. $\frac{7}{10} + \frac{6}{1000}$ = ______
5. $\frac{3}{100} + \frac{3}{1000}$ = ______
6. $\frac{3}{10} + \frac{4}{100}$ = ______
7. $\frac{6}{10} + \frac{6}{100} + \frac{5}{1000}$ = ______
8. $\frac{8}{10} + \frac{8}{1000}$ = ______
9. $\frac{2}{10} + \frac{6}{100} + \frac{3}{1000}$ = ______
10. $\frac{4}{100} + \frac{3}{1000}$ = ______

SCORE /10 0-4 5-8 9-10

TARGETING MATHS HOMEWORK: YEAR 5 © PASCAL PRESS ISBN: 9781925726589

Number & Algebra

AC9M5N01, AC9M5N03

Equivalent Fractions

When you need to add fractions, you often have to change one fraction into another fraction that is equivalent. **Equivalent** means 'has the same value'.

To change $\frac{1}{2}$ into the equivalent in quarters, multiply the numerator and denominator by 2: $\frac{1}{2} = \frac{1 \times 2}{2 \times 2} = \frac{2}{4}$

To change $\frac{1}{3}$ into the equivalent in ninths, multiply the numerator and denominator by 3: $\frac{1}{3} = \frac{1 \times 3}{3 \times 3} = \frac{3}{9}$

Complete the equivalent fractions.

1. $\frac{1}{2} = \frac{1 \times}{2 \times} = \frac{}{16}$
2. $\frac{1}{5} = \frac{1 \times}{5 \times} = \frac{2}{}$
3. $\frac{1}{5} = \frac{1 \times}{5 \times} = \frac{}{15}$
4. $\frac{1}{3} = \frac{1 \times}{3 \times} = \frac{}{12}$
5. $\frac{1}{4} = \frac{1 \times}{4 \times} = \frac{3}{}$
6. $\frac{2}{5} = \frac{2 \times}{5 \times} = \frac{4}{}$
7. $\frac{2}{3} = \frac{2 \times}{3 \times} = \frac{}{6}$
8. $\frac{3}{6} = \frac{3 \times}{6 \times} = \frac{}{18}$
9. $\frac{3}{5} = \frac{\times}{\times} = \frac{9}{}$
10. $\frac{4}{5} = \frac{\times}{\times} = \frac{}{20}$
11. $\frac{4}{7} = \frac{\times}{\times} = \frac{}{42}$
12. $\frac{3}{8} = \frac{\times}{\times} = \frac{18}{}$

SCORE /12 0-5 6-10 11-12

Statistics & Probability

AC9M5ST03

The Vet's Busy Week

Every day, the vet's staff record the number of pets treated at the Vet Clinic.

This table shows the data collected last week.

	Mon.	Tue.	Wed.	Thu.	Fri.	Total
Cats	3	11	13	11	7	
Dogs	3	11	12	6	10	
Kittens	7	7	12	12	10	
Puppies	5	6	12	12	8	
Rabbits	4	5	12	12	7	
Total						

Complete the totals in the table and then write five questions that the data could answer.

1. ______________________________
2. ______________________________
3. ______________________________
4. ______________________________
5. ______________________________

SCORE /5 0-2 3-4 5

TERM 2

AC9M5M02

TERM 2

Finding Areas

A good way to find the area of a shape is to look for the regular shapes that it is made of.

The area of the rectangle in the middle of this shape is found by multiplying the width by the height.

Then the four triangles can be added in as they are each half a square.

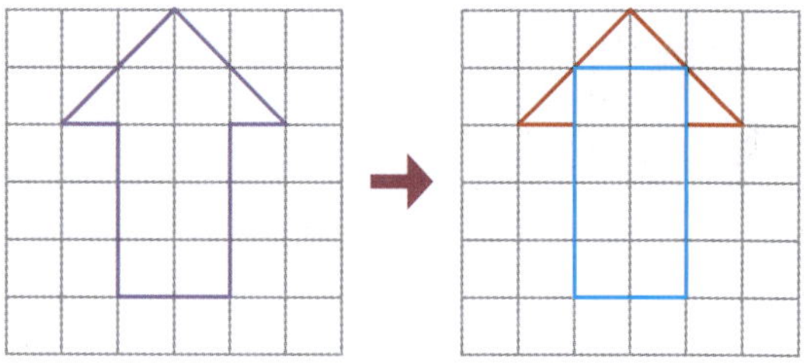

Area of the Shape = $2 \times 4 + \frac{1}{2} + \frac{1}{2} + \frac{1}{2} + \frac{1}{2} = 8 + 2 = 10$ squares

Show how to find the areas of these shapes by drawing on the diagrams (1 point) and then writing a number sentence and the answer (1 point).

1. 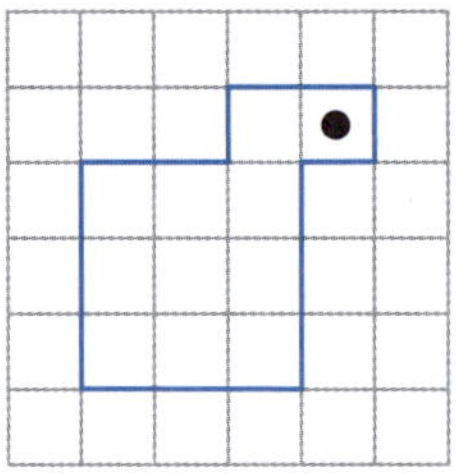

Area
= ____________
= ____________

2. 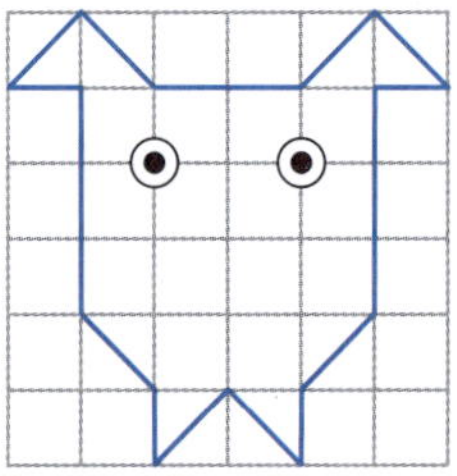

Area
= ____________
= ____________

3. 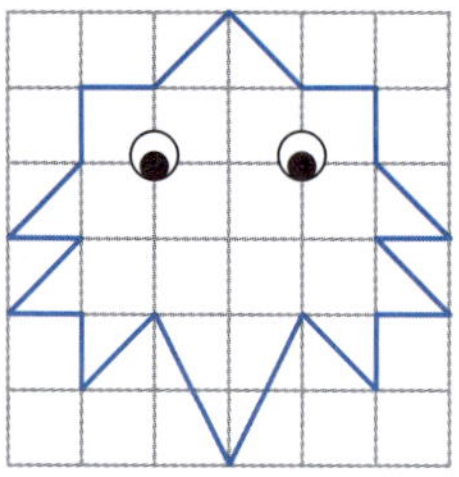

Area
= ____________
= ____________

4. 

Area
= ____________
= ____________

SCORE /4 0-1 2-3 4

NAPLAN*-style Questions

1. What decimal number is the arrow pointing to on this line? 0·

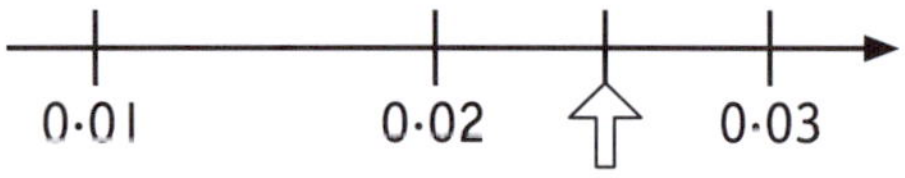

2. Which of these rectangles has the largest fraction of its area shaded?

◯ 
◯ 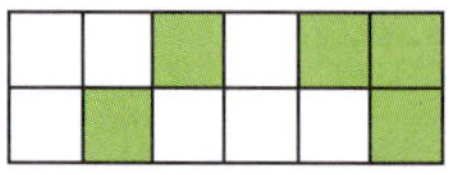
◯ 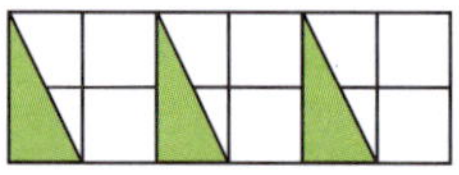
◯ 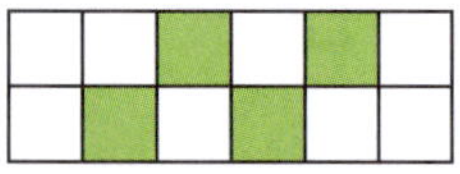

3.  represents one third. Which of these represents one whole?

◯ 
◯ 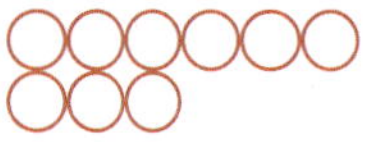
◯ 
◯ 

SCORE /3 0-1 2 3

TARGETING MATHS HOMEWORK: YEAR 5 © PASCAL PRESS ISBN: 9781925726589

Problem of the Week

AC9M5N10

Fundraising Cakes

"Well done indeed, class.

You made three enormous chocolate cakes to raise money for the disaster fund to help our community after the floods. I am very proud of you.

Can you show me how much you sold and what is left over?"

The class were very keen to do that and so they made a list of the sales.

The class had sold:

- $\frac{1}{4}$ of a cake to the first person.
- $\frac{2}{3}$ of a cake to the second person.
- $\frac{1}{6}$ of a cake to the third person.
- $\frac{3}{12}$ of a cake to the fourth person.
- $\frac{1}{4}$ of a cake to the fifth person.
- $\frac{1}{4}$ of a cake to the sixth person.
- $\frac{5}{12}$ of a cake to the seventh person.
- $\frac{1}{4}$ of a cake to the eighth person.
- $\frac{1}{4}$ of a cake to the ninth person.
- $\frac{1}{6}$ of a cake to the tenth person.

The Problem: What fraction of a cake did the class have left over?

To Find the Solution

When we solved this problem in class, Mrs Green suggested that we use three diagrams for the cakes, like this:

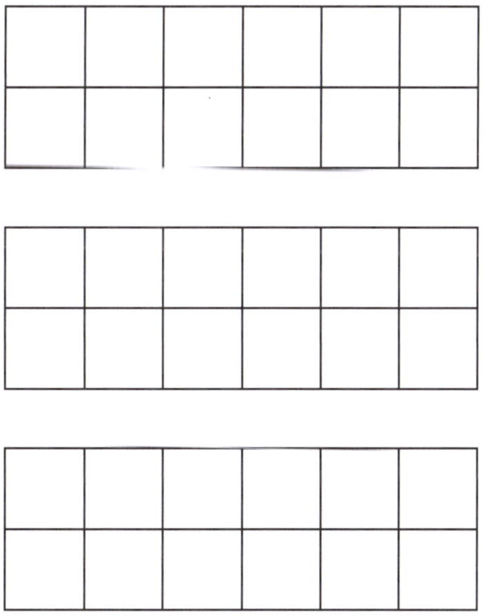

You can use these cake diagrams to help you solve the problem.

And remember what you have learned about equivalent fractions, as that will help too.

The Solution

Mrs Green's class have — of a cake left over.

Review: Unit 9 and Unit 10

AC9M5N01, AC9M5N02, AC9M5N10, AC9M5A02, AC9M5M04

Write the multiple and LCM (lowest common multiple) of each set of factors.

	Factors	Multiple	LCM
(1)	2, 4, 8		
(2)	3, 5, 6		
(3)	6, 5, 4		

Write the decimals marked by the ? on these number lines.

(4) 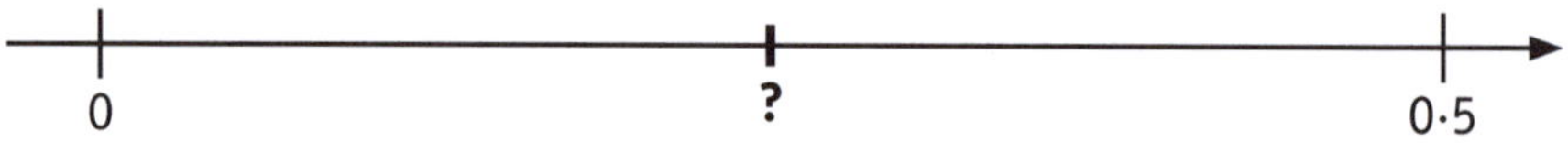

Decimal = ______

(5) 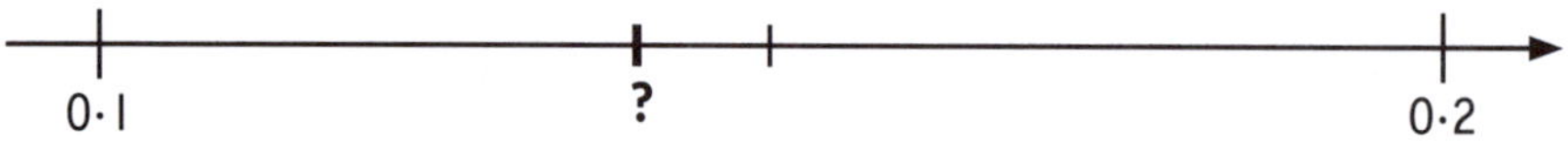

Decimal = ______

(6) 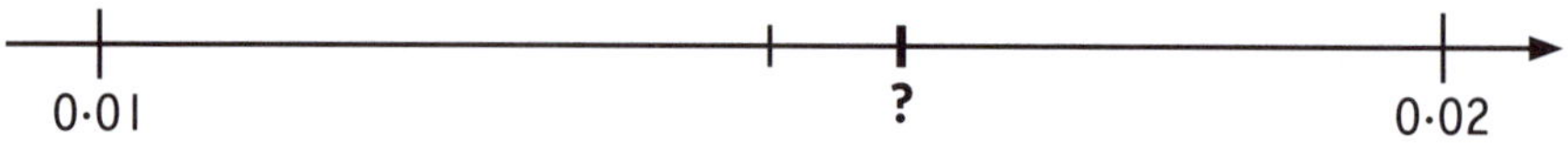

Decimal = ______

Complete these truth sentences using = or ≠.

(7) 8 × 4 ☐ 4 × 2 × 2 × 2

(8) 4 × 9 ☐ 4 × 6 × 3

(9) 24 ÷ 6 ☐ 12 ÷ 3

(10) 36 × 4 ☐ 72 ÷ 2 × 2

(11) 12 × 7 ☐ 28 × 3

(12) 54 ÷ 9 × 2 ☐ 48 ÷ 6 × 2

Estimate how many degrees are in each of these angles and then check with your protractor.

(13) 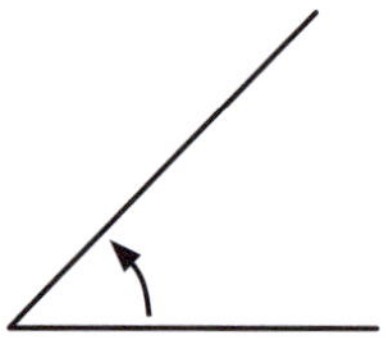

Estimate: ______

Actual: ______

(14) 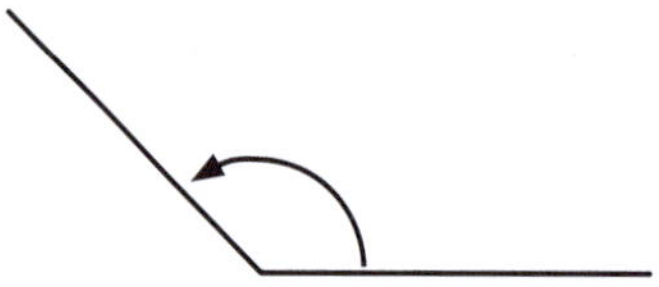

Estimate: ______

Actual: ______

(15) 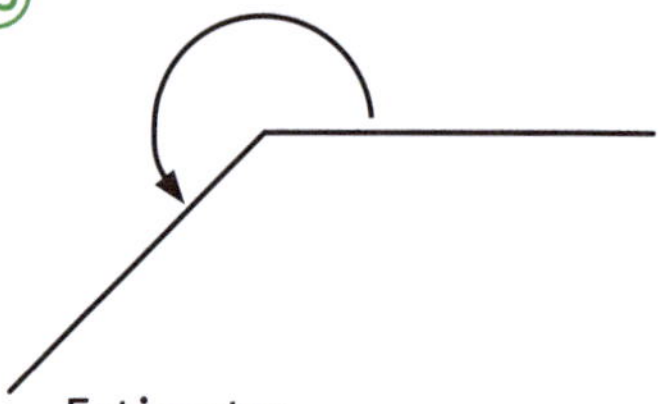

Estimate: ______

Actual: ______

Use factors to solve these multiplications.

		First step	Second step
(16)	For 42 × 4, use ___ = ___ × ___	____ × ___ = _____	____ × ___ = _____
(17)	For 42 × 6, use ___ = ___ × ___	____ × ___ = _____	____ × ___ = _____
(18)	For 42 × 9, use ___ = ___ × ___	____ × ___ = _____	____ × ___ = _____

SCORE /18 (0-8) (9-16) (17-18)

TARGETING MATHS HOMEWORK: YEAR 5 © PASCAL PRESS ISBN: 9781925726589

Review: Unit 11 and Unit 12

TERM 2 REVIEW

AC9M5N03, AC9M5N04, AC9M5N05, AC9M5ST02, AC9M5P01

Shade each fraction and write an equivalent fraction.

1 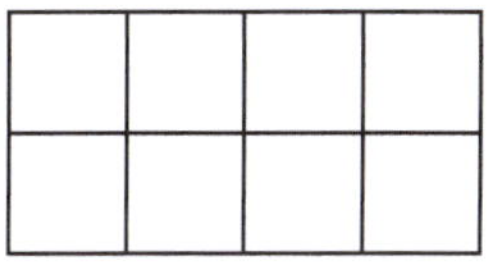

$\frac{4}{8}$ or —

2 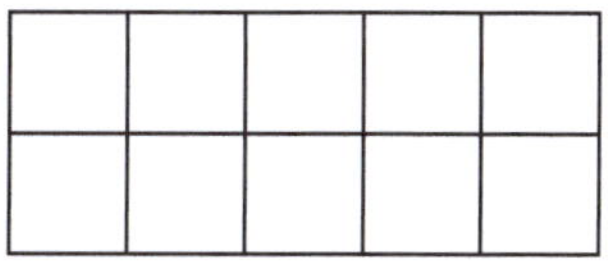

$\frac{4}{10}$ or —

3 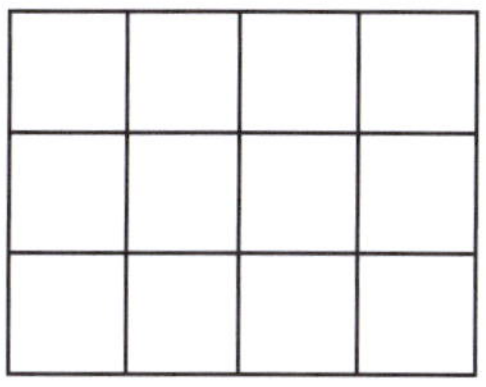

$\frac{4}{12}$ or —

What fraction of these shapes is unshaded? Write your answer as a percentage.

4 Unshaded = ___%

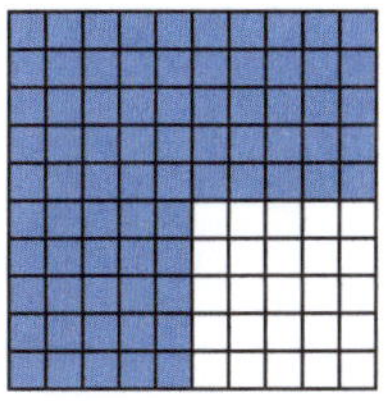

5 Unshaded = ___%

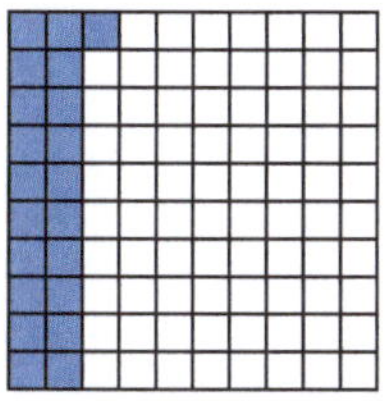

6 Unshaded = ___%

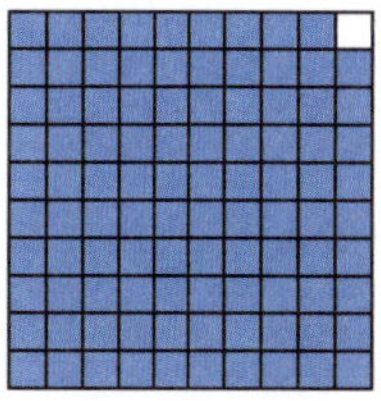

This graph shows the data for tram rides for two days last week.

7 Which day was busier? ____________________

8 How many people were on the tram at 11:00 on Monday?

_____

9 How many more people were on the tram at 11:00 on Tuesday than on Monday?

_____

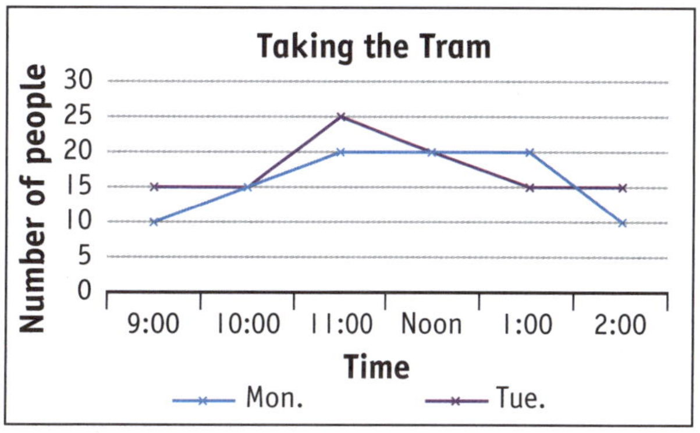

Complete these fraction additions and subtractions.

10 $\frac{3}{8} + \frac{4}{8} =$ —

11 $\frac{1}{3} + \frac{1}{6} =$ —

12 $\frac{1}{2} - \frac{3}{8} =$ —

13 $\frac{3}{5} - \frac{2}{10} =$ —

14 $\frac{2}{3} - \frac{1}{2} =$ —

15 $\frac{3}{9} - \frac{1}{3} =$ —

Show the chance of spinning blue on the probability lines.

16 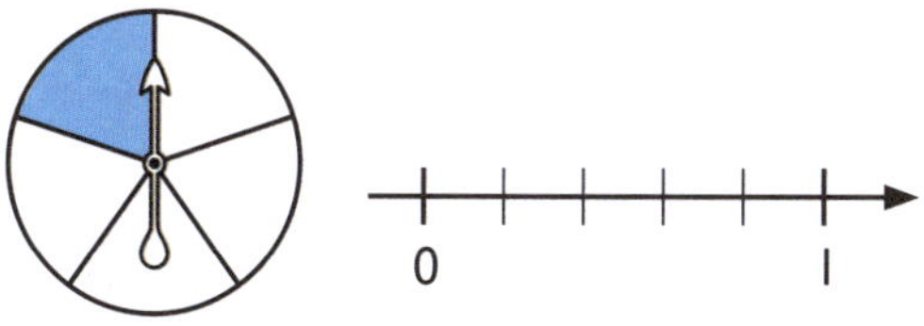

17 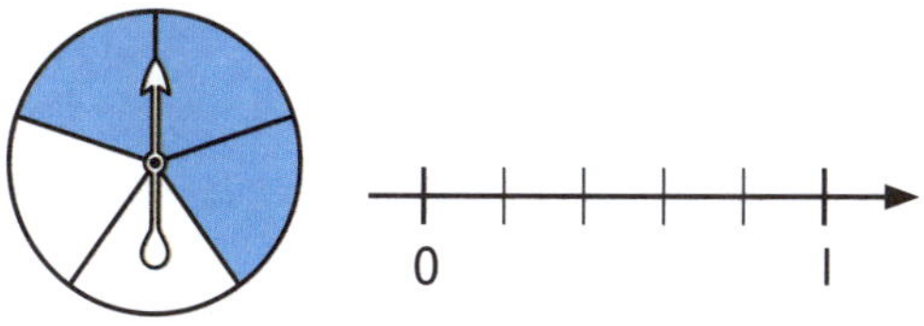

18 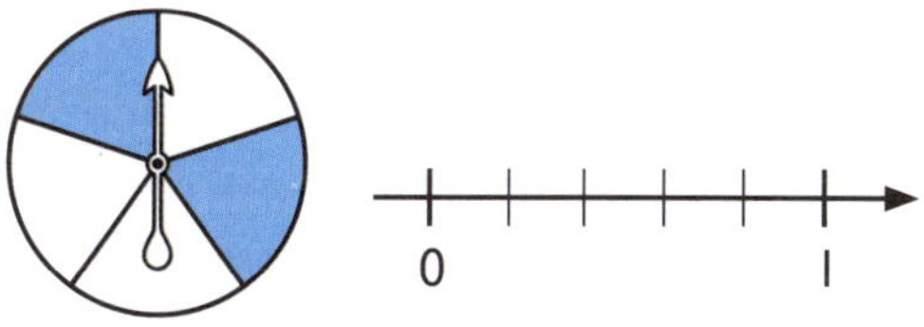

19 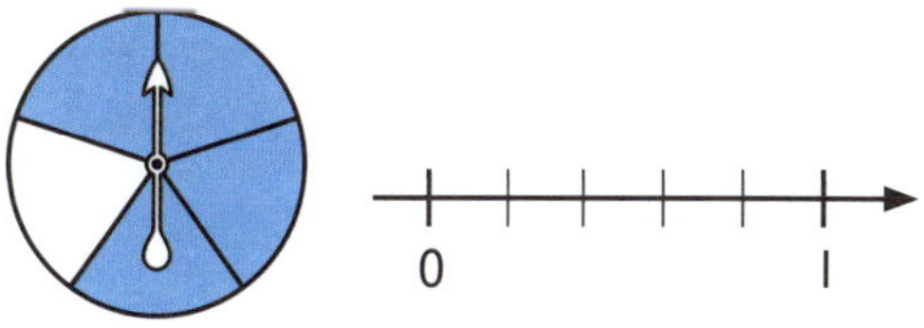

SCORE /19 0-9 10-17 18-19

Review: Unit 13 and Unit 14

AC9M5N07, AC9M5N08, AC9M5M03, AC9M5ST01, AC9M5ST03

Complete these divisions with remainders.

1. 26 ÷ 5 = ___ r ___
2. 39 ÷ 4 = ___ r ___
3. 30 ÷ 7 = ___ r ___
4. 58 ÷ 8 = ___ r ___

Answer these division questions and include the fraction part each time.

5. 7 metres of ribbon will be made into 4 identical bows.
   How much ribbon will be used for each bow?

   ___________ m

6. Asha is dividing 13 kilograms of tomatoes into 6 boxes.
   What mass of tomatoes will be in each box?

   ___________ kg

This graph of favourite sports is incomplete.
Complete the chart to show that:

7. 10% preferred hockey.
8. Twice as many preferred netball to hockey.
9. The same percentage preferred football as cricket.

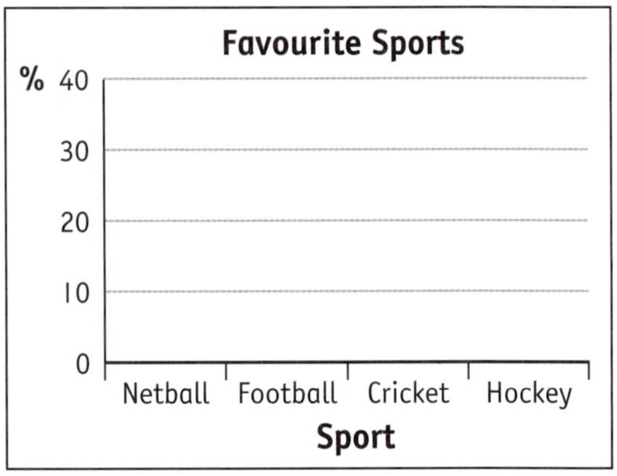

Use round and adjust in your head to answer these questions.

10. 4 at $3·95 each = $__________
11. 6 at $1·80 each = $__________
12. 7 at $1·15 each = $__________
13. 3 at $15·25 each = $__________

What is the mode for these dice strings?

14. For 4, 3, 3, 4, 2, 6, 5, 1, 4, the mode is ____.
15. For 5, 6, 6, 3, 2, 1, 5, 6, 3, the mode is ____.

Use these timelines to help you answer the following questions.

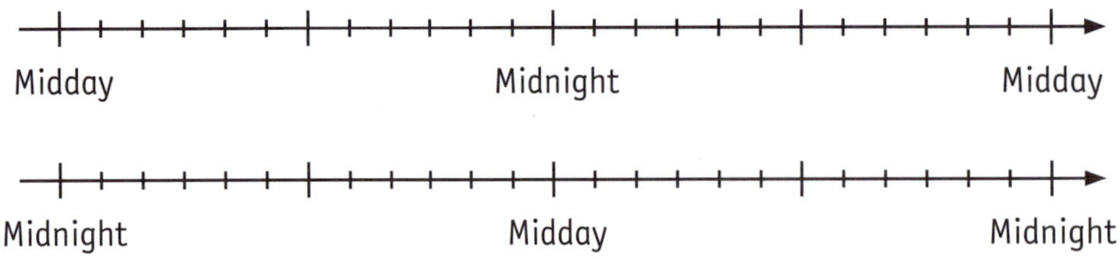

If London is 10 hours behind Melbourne, what time will it be in London at these Melbourne times? Answer in am/pm form.

16. 14:30 in Melbourne will be __________ in London.
17. 8:30 am in Melbourne will be __________ in London.
18. 11:20 pm in Melbourne will be __________ in London.

SCORE /18 0-8 9-16 17-18

TARGETING MATHS HOMEWORK: YEAR 5 © PASCAL PRESS ISBN: 9781925726589

Review: Unit 15 and Unit 16

TERM 2 REVIEW

AC9M5N01, AC9M5N03, AC9M5A01, AC9M5A02, AC9M5SP02, AC9M5ST03

Complete these equations by finding the unknowns.

1. $3 \times \square = \square \div 2 = 27$
2. $6 \times \square = \square \div 2 = 42$
3. $3 \times \square = 72 \div \square = 18$
4. $\square \times 4 = 96 \div \square = 48$

What are the proper factors of these numbers?

5. 32: ___ ___ ___ ___
6. 44: ___ ___ ___ ___

Write the coordinates for each of these symbols.

7. ▲ is at (__, __).
8. ⬟ is at (__, __).
9. ⬢ is at (__, __).
10. ◆ is at (__, __).
11. ✖ is at (__, __).
12. ★ is at (__, __).

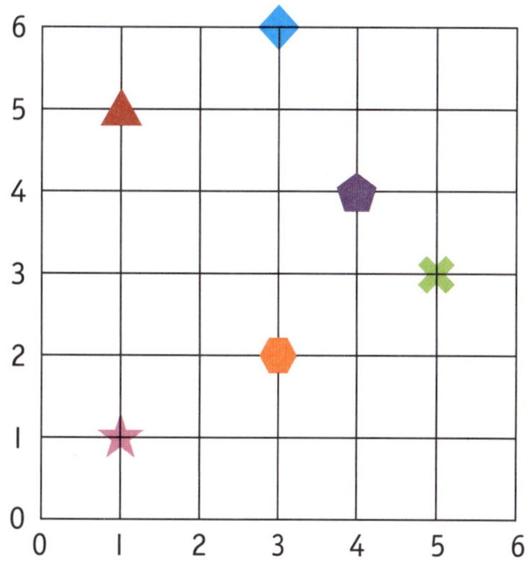

Combine these fractions and write them as decimals.

13. $\frac{3}{10} + \frac{4}{100} = 0\cdot$ ________
14. $\frac{4}{10} + \frac{3}{100} + \frac{4}{1000} = 0\cdot$ ________
15. $\frac{4}{100} + \frac{3}{1000} = 0\cdot$ ________
16. $\frac{5}{10} + \frac{5}{1000} = 0\cdot$ ________

Complete the equivalent fractions.

17. $\frac{1}{2} = \frac{3}{}$
18. $\frac{4}{5} = \frac{}{10}$
19. $\frac{4}{16} = \frac{}{4}$
20. $\frac{8}{12} = \frac{2}{}$

Use the table to answer the questions about the vet's busy week.

21. Complete the total columns.
22. How many puppies were treated?

    ____
23. On which day were 40 pets treated?

    ____
24. The vet is thinking of closing for half a day each week.
    Which day would you recommend?

    ____
25. Why do you think that?

    ____

Animals	Mon.	Tue.	Wed.	Thu.	Fri.	Total
Cats	13	8	10	5	5	
Dogs	12	10	12	10	9	
Kittens	12	8	13	5	5	
Puppies	4	5	13	12	6	
Rabbits	9	9	14	9	7	
Total						

SCORE /25 0-12 13-23 24-25

AC9M5N01

Ordering Decimal Fractions

To arrange a string of fractions and decimals in order from smallest to largest, you need to convert them all to either fractions or decimals.

For $1\frac{6}{10}$, $\frac{150}{100}$, 0·95, $\frac{35}{10}$,

we could write them all as decimals: 1·6, 1·5, 0·95, 3·5

or all as fractions: $\frac{160}{100}$, $\frac{150}{100}$, $\frac{95}{100}$, $\frac{350}{100}$

and then arrange them in order from smallest to largest:

0·95, 1·5, 1·6, 3·5 or $\frac{95}{100}$, $\frac{150}{100}$, $\frac{160}{100}$, $\frac{350}{100}$.

Convert and then arrange these fraction/decimal strings in order from smallest to largest.

1. 0·99, $\frac{56}{100}$, $\frac{9}{10}$, $\frac{13}{10}$

   Decimals: ____________ Fractions: ____________

   Order in decimals from smallest to largest: ____________

2. $\frac{36}{100}$, $\frac{99}{100}$, 2·6, $\frac{18}{10}$

   Decimals: ____________ Fractions: ____________

   Order in decimals from smallest to largest: ____________

3. $\frac{3}{10}$, $\frac{13}{100}$, 1·3, $\frac{135}{100}$

   Decimals: ____________ Fractions: ____________

   Order in decimals from smallest to largest: ____________

4. $\frac{8}{10}$, $\frac{18}{100}$, 0·76, $\frac{270}{1000}$

   Decimals: ____________ Fractions: ____________

   Order in decimals from smallest to largest: ____________

5. 0·009, $\frac{9}{10}$, $\frac{4}{1000}$, 0·45

   Decimals: ____________ Fractions: ____________

   Order in decimals from smallest to largest: ____________

SCORE /5 (0-2) (3-4) (5)

TERM 3

Quick Check

Write these fractions as decimals.

1. $\frac{27}{100}$ ___·___
2. $\frac{34}{10}$ ___·___
3. $\frac{305}{100}$ ___·___
4. $\frac{46}{1000}$ ___·___
5. $\frac{5}{10}$ ___·___

Write these decimals as fractions.

6. 0·63 ______
7. 0·036 ______
8. 1·35 ______
9. 3·006 ______
10. 0·065 ______

SCORE /10 (0-4) (5-8) (9-10)

TARGETING MATHS HOMEWORK: YEAR 5 © PASCAL PRESS ISBN: 9781925726589

AC9M5N02

Rules for Divisibility

The rules for divisibility tell you if a large number can be **divided evenly** (that means **with no remainder**) by a small number. If you know the rules for divisibility, you can quickly tell if a large number is divisible by a small number.

You already know these rules:
- A number is divisible by **2** if it is even.
- A number is divisible by **5** if it ends in 5 or 0.

But do you know these rules?
- A number is divisible by **3** if its digits add to a multiple of 3.
- A number is divisible by **6** if it is even and divisible by 3.
- A number is divisible by **4** if the last two digits are a multiple of 4.
- A number whose **hundreds are even** is divisible by **8** if the last two digits are a multiple of 8. If the **hundreds are odd**, the last two digits must be a multiple of 4 but not of 8.
- A number is divisible by **9** if its digits add to a multiple of 9.

Use ticks and crosses to complete the divisibility tables.

	The number	is divisible by						
		2	3	4	5	6	8	9
1	69 870	✓	✓	✗				
2	34 264							
3	72 369							
4	12 645							
5	35 786							

	The number	is divisible by						
		2	3	4	5	6	8	9
6	45 607							
7	93 645							
8	84 279							
9	65 348							
10	8 658 345							

SCORE /10 0-4 5-8 9-10

Statistics & Probability

AC9M5P02

10 Counters in the Bag

I have made up a bag of 10 counters. To make a **play**, without looking I take one counter out, record its colour and then put it back in the bag and give the bag a shake. These are the colours from the first ten plays that I made:

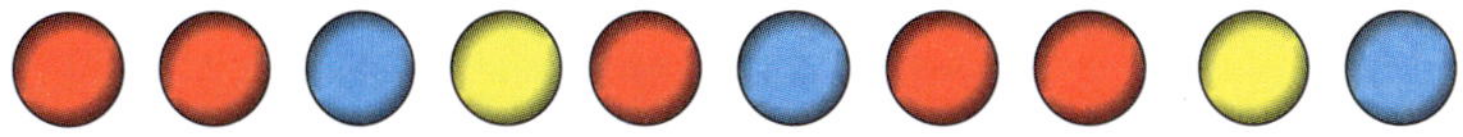

Use the chance terms Certain, Very likely, Likely, Unlikely or Impossible to describe the likelihood of the following outcomes for Ella's next ten plays.

1. There are no green counters. ______________
2. Most of the counters are red. ______________
3. There are only two yellow counters. ______________
4. There are the same number of red counters as the other colours combined. ______________
5. There is only one blue counter. ______________

SCORE /5 0-2 3-4 5

AC9M5SP01

Folding Nets

When the first five of these nets are folded, only one will make a cube with a continuous snake, from its head to the tip of its tail. Which net is that?

**I left you with a blank net for number 6.
Draw in your own snake that will be continuous from its head to the tip of its tail when the net is folded up.**

1

4
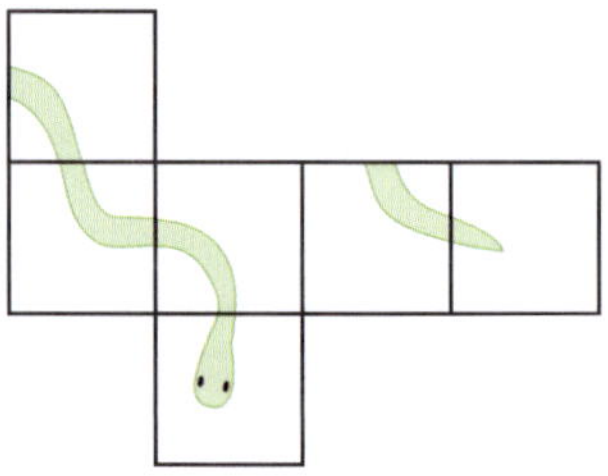

2
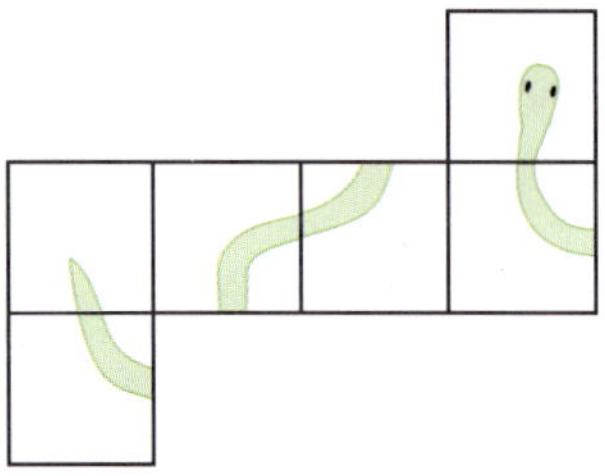

5
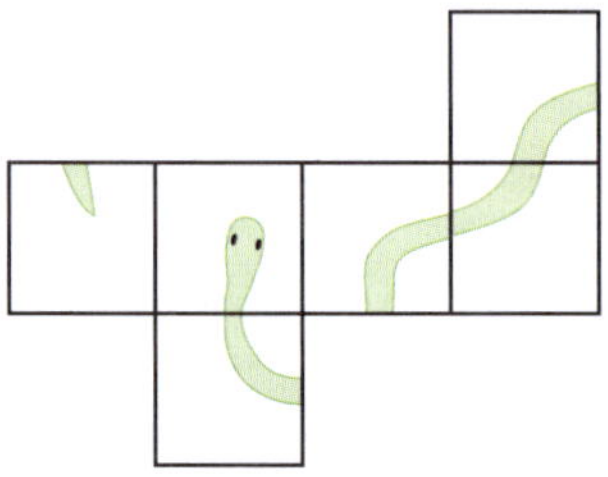

3
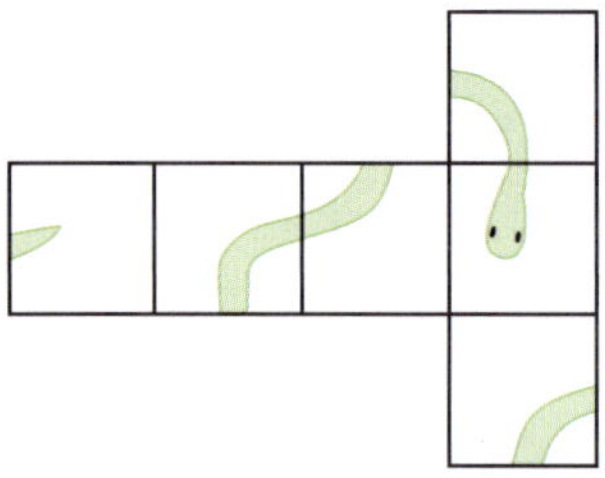

6
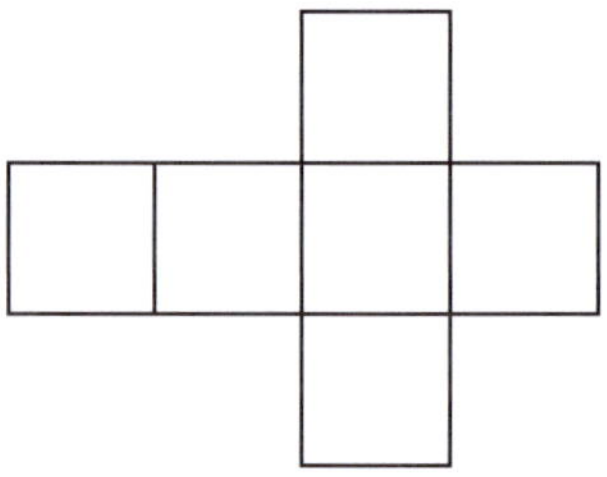

SCORE /6 0-2 3-4 5-6

NAPLAN*-style Questions

1. What number is $\frac{1}{100}$ less than 2·343?

   ◯ 2·342 ◯ 2·243 ◯ 2·333 ◯ 2·343

2. Which of these numbers is divisible by 3?

   ◯ 9126 ◯ 7343 ◯ 8243 ◯ 7315

3. Which of these nets will fold up into a triangular pyramid?

   ◯ 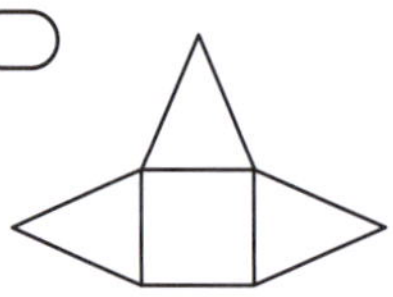 ◯ 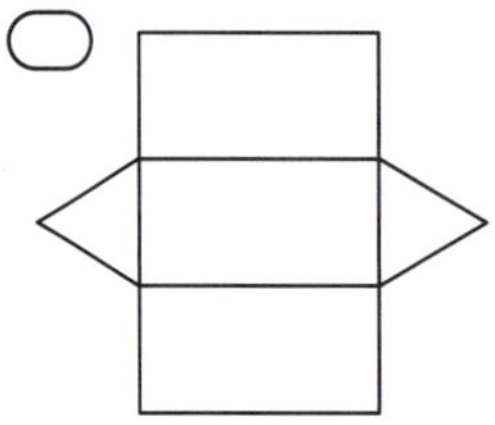 ◯ 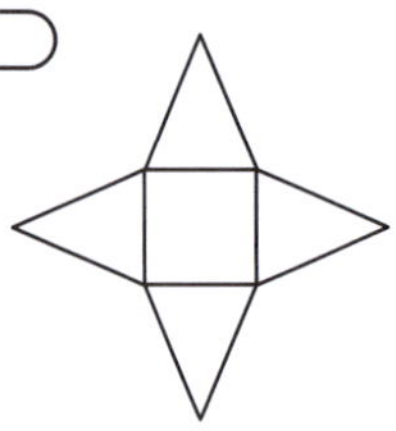 ◯ 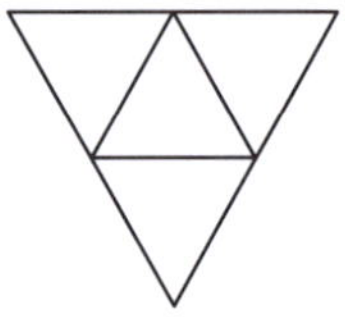

SCORE /3 0-1 2 3

TARGETING MATHS HOMEWORK: YEAR 5 © PASCAL PRESS ISBN: 9781925726589

Problem of the Week

AC9M5N08

Making Multiples

Making Multiples is a game that uses only the numbers 1–9 from a pack of cards.

The cards are shuffled, and players are each dealt five cards.

Each player arranges their cards to make a 5-digit number that is a multiple of 2, 3, 4, 5, 6, 8 or 9.

The score for each round is the total of the numbers that the 5-digit number is a multiple of.

For example:

12 345 is a multiple of 3 and 5 and so scores 8.

53 412 is a multiple of 2, 3, 4 and 6 and so scores 15.

13 452 is a multiple of 2, 3 and 6 and so scores 11.

51 432 is a multiple of 2, 3, 4, 6 and 8. This arrangement scores 23 – WOW!

Whose cards can score the most?

These are the cards dealt to Asha, Ravi, Sam and Leo.
Whose cards can be arranged to make the highest score?

① 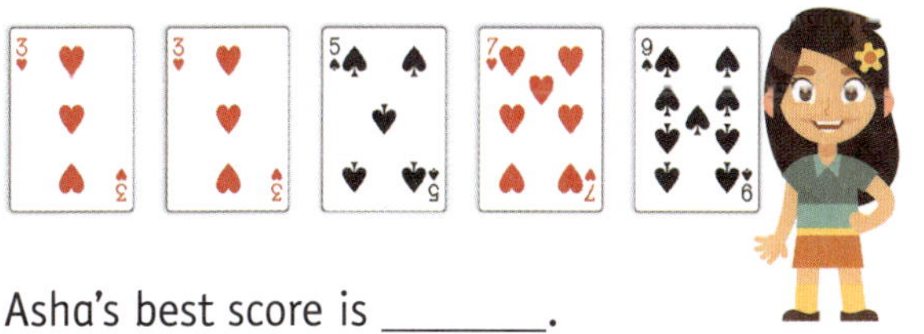

Asha's best score is ______.

③ 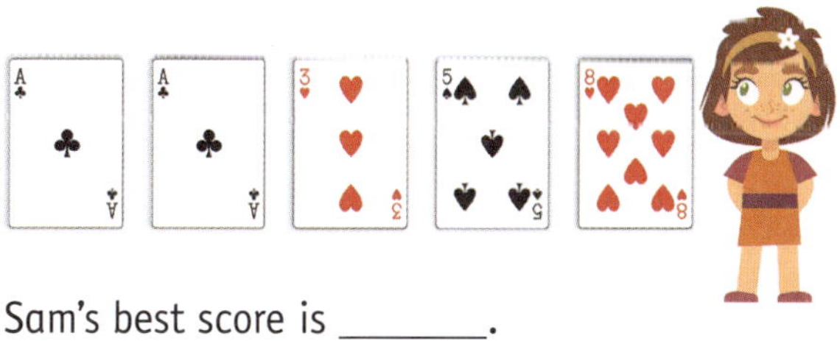

Sam's best score is ______.

② 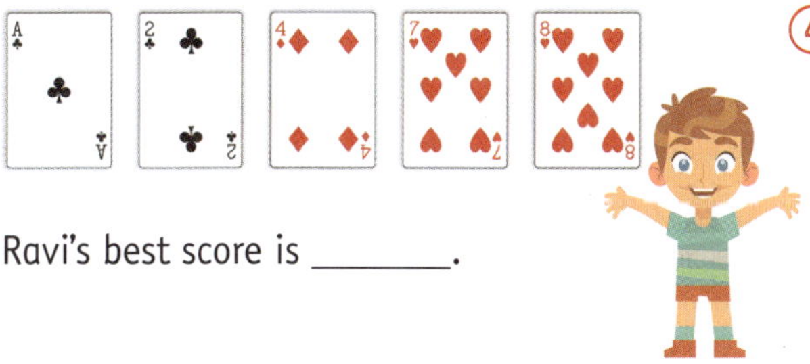

Ravi's best score is ______.

④ 

Leo's best score is ______.

⑤ **The person with the highest score is ____________.**

TERM 3

Percentages in Daily Events

Our phones, tablets and other devices all display percentages.

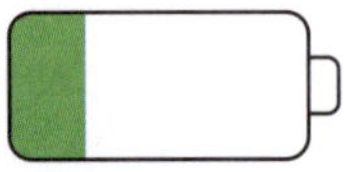

This is a common symbol used to let us know how much battery life our device has left.

When the symbol looks like this, we know that our battery is 100% charged and will last quite a long time.

When the symbol looks like this, we know that our battery is flat. Time to plug the device in and start charging the battery.

What percentage of battery remains?

1. 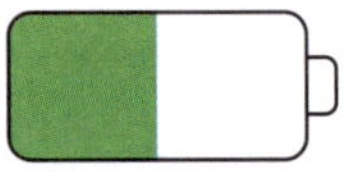

   ___% remains

2. 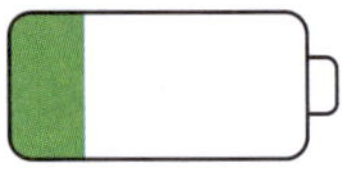

   ___% remains

3. 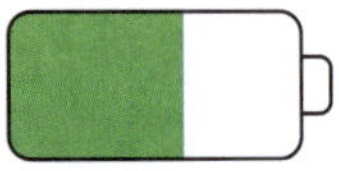

   ___% remains

4. 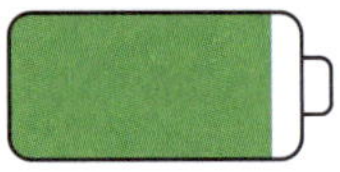

   ___% remains

5. 

   ___% remains

6. 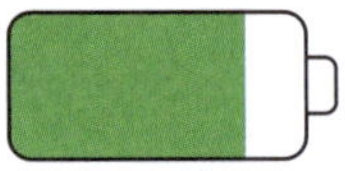

   ___% remains

Shade the batteries to show these percentages.

7. 10% remains

8. 40% remains

   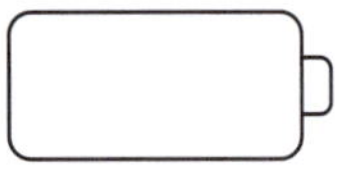

9. 5% remains

If 30% of the battery remains, it means that 70% has been used because 30% + 70% = 100%.

What percentage has been used?

10. If 15% remains, ___% has been used.
11. If 25% remains, ___% has been used.
12. If 95% remains, ___% has been used.

SCORE /12 0-5 6-10 11-12

Quick Check

How much battery remains after this % has been used?

1. 23% ______
2. 94% ______
3. 87% ______
4. 78% ______
5. 26% ______

How much battery has been used if this % remains?

6. 16% ______
7. 50% ______
8. 75% ______
9. 68% ______
10. 32% ______

SCORE /10 0-4 5-8 9-10

TARGETING MATHS HOMEWORK: YEAR 5 © PASCAL PRESS ISBN: 9781925726589

Number & Algebra

AC9M5N03

Fractions of a Shape

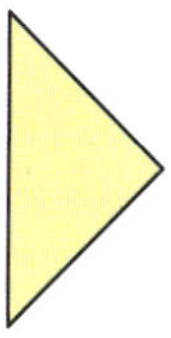

If this triangle is $\frac{1}{3}$ of a shape, what fraction do each of these shapes represent? Write a mixed number where appropriate.

① ______ 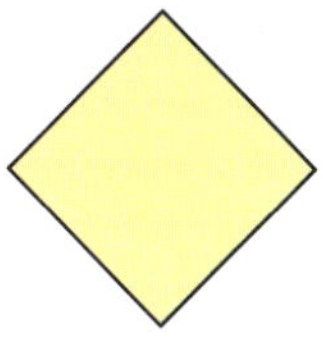

② ______ 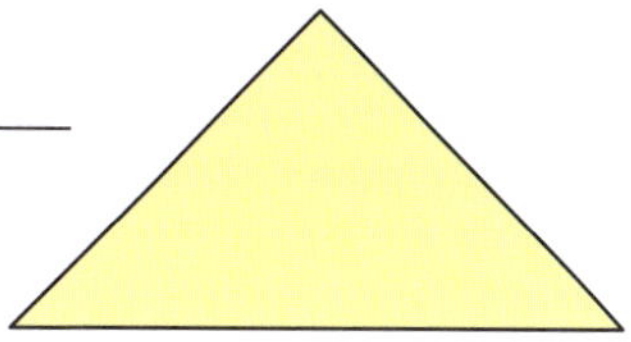

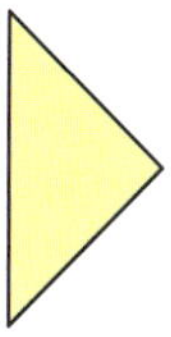

If this triangle is $\frac{3}{4}$ of a shape, what fraction do each of these shapes represent? Write a mixed number where appropriate.

③ ______ 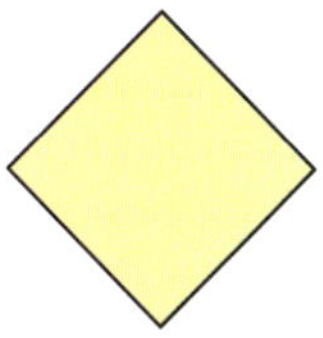

④ ______ 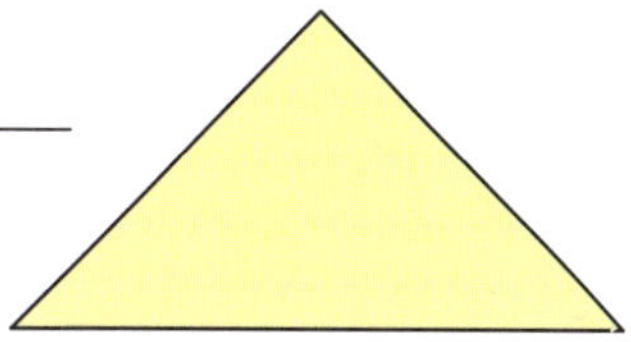

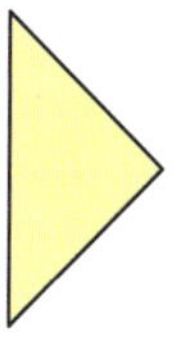

If this triangle is $\frac{1}{5}$ of a shape, what fraction do each of these shapes represent? Write a mixed number where appropriate.

⑤ ______ 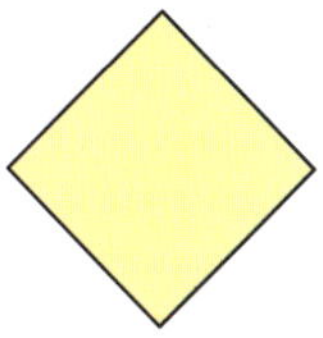

⑥ ______ 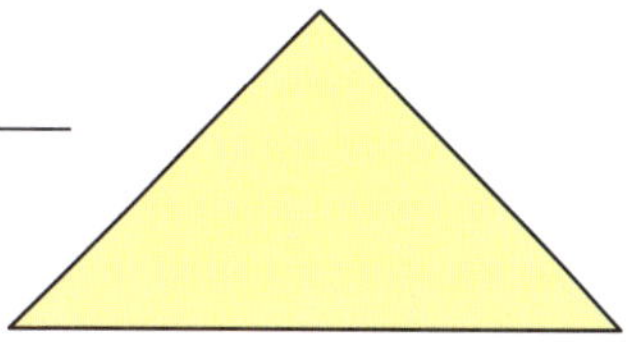

SCORE /6 0-2 3-4 5-6

Statistics & Probability

AC9M5ST01

Broken Bones

The radiographer used a spreadsheet to make this chart of how many broken bones had been x-rayed in the last 10 days.

The radiographer then had to include the following information in her report.

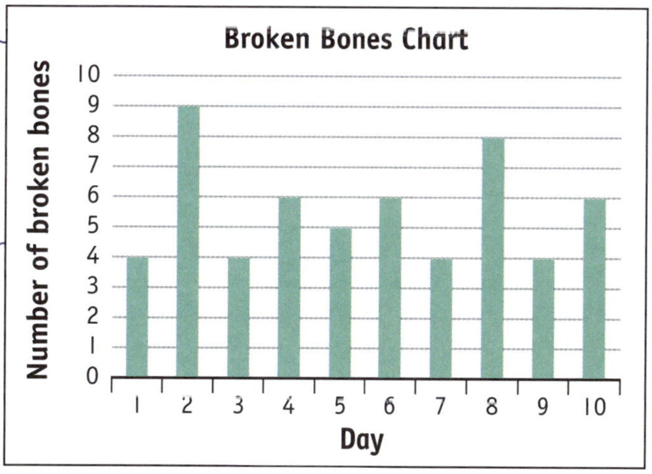

① How many broken bones altogether? ______

② Which two days were the busiest? ______

③ What was the mode of the data? ______

④ Which was busier: the first 5 days or the second 5 days? ______

⑤ A quarter of the broken bones were arms. How many broken arms were x-rayed? ______

SCORE /5 0-2 3-4 5

AC9M5M01

Units of Measurement

These are the units of measurement that we use for **mass**, **capacity** and **length**:
millimetre, kilogram, litre, centimetre, milligram, gram, metre, millilitre, kilometre.

Write the units in the correct category and in order from smallest to largest.

1. Mass: __________ __________ __________
2. Capacity: __________ __________
3. Length: __________ __________ __________ __________

What might you be measuring if you got these measurements?

4. 50 km __________
5. 25 cm __________
6. 10 mL __________
7. 2 milligrams __________
8. 2 L __________
9. 1455 mm __________
10. $\frac{1}{2}$ kg __________
11. $\frac{1}{4}$ kg __________
12. $\frac{1}{8}$ kg __________

SCORE /12 0-5 6-10 11-12

TERM 3

NAPLAN*-style Questions

1.  How much energy is left in this battery?

◯ 20% ◯ 100% ◯ 80% ◯ 40%

2. 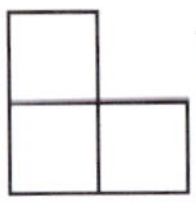 This shape represents $\frac{1}{3}$.

What fraction does this shape represent?

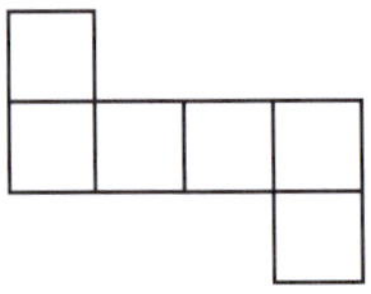

◯ $\frac{6}{3}$ ◯ $\frac{1}{6}$ ◯ $1\frac{1}{3}$ ◯ $\frac{2}{3}$

3. The chart shows how many scoring jump shots these basketball players made in practice last Wednesday.

Ravi netted 6 jump shots.

How many more jump shots did Will net than Asha?

☐

SCORE /3 0-1 2 3

TARGETING MATHS HOMEWORK: YEAR 5 © PASCAL PRESS ISBN: 9781925726589

Problem of the Week

AC9M5M01

The Jeweller's Scales

The jeweller has some very accurate scales for weighing small amounts of precious metals.

She only has three weights that she can use with the scales:

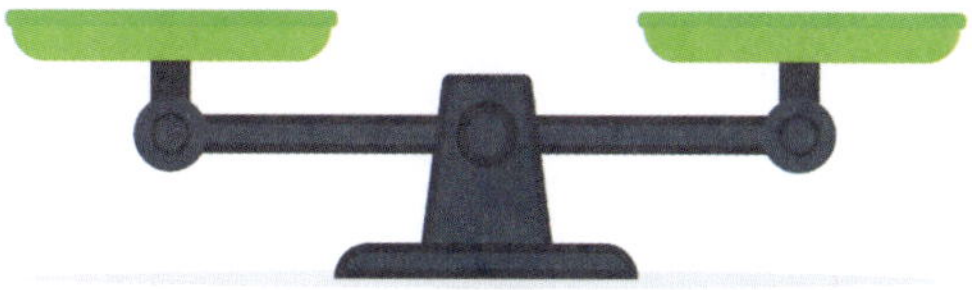

Which weights should the jeweller use to weigh pieces of metal with these masses?

- On Side 1, she puts her weights.
- On Side 2, she puts the metal piece and, if needed to make the balance work, one of her weights.

Mass	Side 1 Weight(s)	Side 2: Metal piece + weight(s)
4 g		
7 g		
8 g		
9 g		
2 g		

The Challenge

What different masses can be weighed on the jeweller's scales using these weights?

With these weights, the largest mass that can be balanced is 1 + 3 + 6 + 13 = 23 g.

Use the tables to show how each mass from 1 g to 23 g can be balanced.

- Draw a ✓ to put that weigth on Side 1.
- Draw a ✗ to put that weight on Side 2 with the metal piece being weighed.

Mass	Weights			
	1 g	3 g	6 g	13 g
1 g				
2 g				
3 g				
4 g				
5 g				
6 g				
7 g				
8 g				

Mass	Weights			
	1 g	3 g	6 g	13 g
9				
10				
11				
12				
13				
14				
15				
16				

Mass	Weights			
	1 g	3 g	6 g	13 g
17				
18				
19				
20				
21				
22				
23				

TERM 3

AC9M5N05

Adding Fractions

Using equivalent fractions is an important strategy for the addition of fractions.
To add $\frac{1}{3}$ and $\frac{1}{6}$, you can change $\frac{1}{3}$ to its equivalent form of $\frac{2}{6}$. Then add.
Usually, the answer to an addition of fractions is written in its simplest form:
$\frac{1}{3} + \frac{1}{6} = \frac{2}{6} + \frac{1}{6} = \frac{3}{6} = \frac{1}{2}$

Use equivalent fractions to add. Remember to give the answer in its simplest form.

1. $\frac{3}{8} + \frac{1}{4} =$ — + — = —
2. $\frac{3}{4} + \frac{5}{8} =$ — + — = — = —
3. $\frac{2}{3} + \frac{1}{6} =$ — + — = —
4. $\frac{1}{5} + \frac{3}{10} =$ — + — = — = —
5. $\frac{1}{8} + \frac{1}{16} =$ — + — = —
6. $\frac{1}{2} + \frac{1}{8} =$ — + — = —
7. $\frac{1}{12} + \frac{1}{4} =$ — + — = — = —
8. $\frac{5}{6} + \frac{1}{12} =$ — + — = —
9. $\frac{3}{8} + \frac{3}{4} =$ — + — = — = —
10. $\frac{3}{5} + \frac{7}{10} =$ — + — = — = —
11. $\frac{4}{5} + \frac{7}{10} =$ — + — = — = —
12. $\frac{5}{6} + \frac{6}{12} =$ — + — = — = —
13. $\frac{7}{8} + \frac{5}{8} =$ — + — = — = —
14. $\frac{27}{10} + \frac{7}{5} =$ — + — = — = —

SCORE /14 (0-6) (7-12) (13-14)

TERM 3

Quick Check

Write these fractions in their simplest form or as mixed fractions.

1. $\frac{6}{8} =$ —
2. $\frac{3}{6} =$ —
3. $\frac{4}{10} =$ —
4. $\frac{8}{4} =$ ___
5. $\frac{9}{6} =$ —
6. $\frac{6}{12} =$ —
7. $\frac{4}{3} =$ —
8. $\frac{8}{6} =$ —
9. $\frac{11}{10} =$ —
10. $\frac{8}{16} =$ —

SCORE /10 (0-4) (5-8) (9-10)

TARGETING MATHS HOMEWORK: YEAR 5 © PASCAL PRESS ISBN: 9781925726589

Number & Algebra

AC9M5N05

Subtracting Fractions

We can also use **equivalent fractions** for the subtraction of fractions.

To subtract $\frac{5}{8}$ from $\frac{3}{4}$ you can change $\frac{3}{4}$ to its equivalent form of $\frac{6}{8}$. Then subtract.

As with addition, the answer to a subtraction of fractions is written in its simplest form. But there is no change needed in this case.

$\frac{3}{4} - \frac{5}{8} = \frac{6}{8} - \frac{5}{8} = \frac{1}{8}$

Use equivalent fractions to subtract.

1. $\frac{1}{2} - \frac{1}{8} = \_\_ - \_\_ = \_\_$
2. $\frac{5}{8} - \frac{1}{4} = \_\_ - \_\_ = \_\_$
3. $\frac{1}{3} - \frac{1}{6} = \_\_ - \_\_ = \_\_$
4. $\frac{3}{4} - \frac{3}{8} = \_\_ - \_\_ = \_\_$
5. $\frac{3}{4} - \frac{1}{8} = \_\_ - \_\_ = \_\_$
6. $\frac{1}{2} - \frac{2}{6} = \_\_ - \_\_ = \_\_$
7. $\frac{3}{6} - \frac{1}{12} = \_\_ - \_\_ = \_\_$
8. $\frac{3}{8} - \frac{1}{16} = \_\_ - \_\_ = \_\_$
9. $\frac{4}{5} - \frac{1}{10} = \_\_ - \_\_ = \_\_$
10. $\frac{7}{8} - \frac{3}{4} = \_\_ - \_\_ = \_\_$

SCORE /10 0-4 5-8 9-10

Statistics & Probability

AC9M5ST01

Fitness Program Table

The local gym has six fitness activities that can be made into a one-hour exercise program.
The personal trainer wants a different workout each weekday.
Each workout should lasts for exactly one hour.
Creating a table can help you figure out possible workout programs.

Show which activities could be included each day to make up 5 different workouts.

Activities	Minutes	Day 1	Day 2	Day 3	Day 4	Day 5
Walking machine	20					
Weights	10					
Pilates	30					
Swimming	25					
Rowing machine	15					
Stretches	5					
Fitness program total						

SCORE /5 0-2 3-4 5

TERM 3

Measurement & Space

AC9M5M02

The Landscape Garden Plan

Our family is giving the back garden a make-over.

We all have to make a plan to show what we want the new garden to look like.

This is my design. It includes a gazebo where we can keep cool in the hot summer and a fishpond because I really like watching fish swim around – maybe I could join them occasionally!

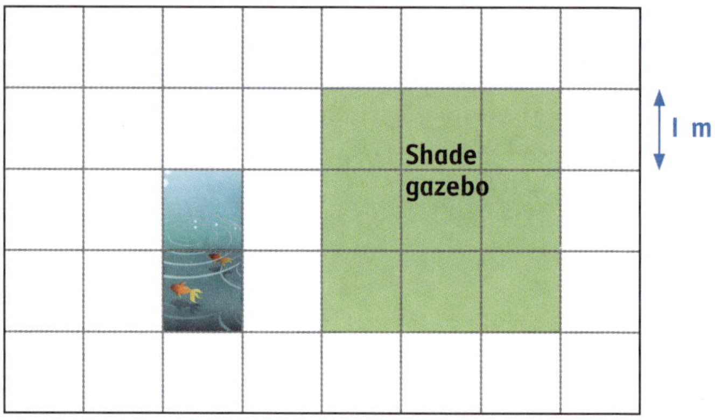

1. How much boundary fencing will be needed? ______ m
2. What is the area of the paving under the gazebo? ______ m^2
3. The fishpond needs a low fence. What is the perimeter of the fishpond? ______ m
4. The rest of the garden needs to be turfed. What area of turf is needed? ______ m^2
5. A bush is to be planted at each of the boundary corners and at every metre around the boundary. How many bushes will be needed?

   _____ bushes
6. A fruit tree will be planted at every point that is 1 metre away from the fence, fishpond and gazebo.
   How many fruit trees are needed?

   _____ fruit trees

SCORE /6 0-2 3-4 5-6

NAPLAN*-style Questions

1. What is the fraction that is marked on the fraction number line?

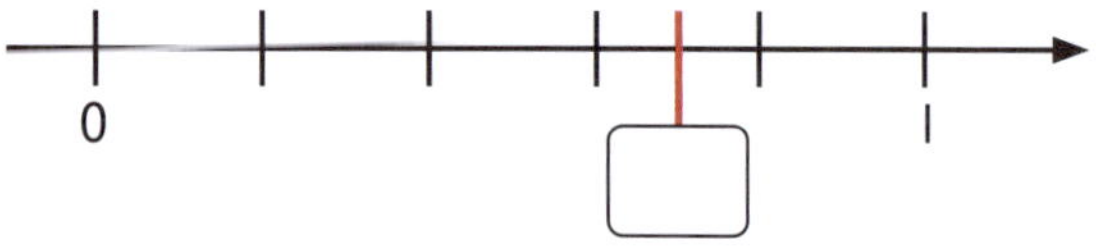

2. $\frac{3}{8}$ of an apple pie has been eaten. What fraction of the pie remains?

   — of the pie

3. This is a diagram of a garden plot.
   What is the perimeter of the plot?

   _____ metres

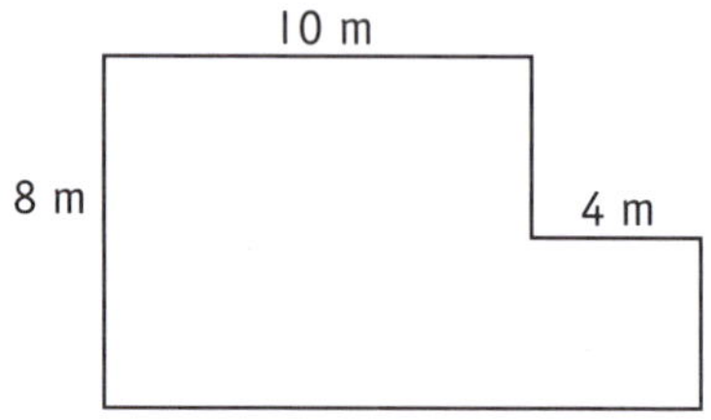

SCORE /3 0-1 2 3

TARGETING MATHS HOMEWORK: YEAR 5 © PASCAL PRESS ISBN: 9781925726589

Baby Animal Petting Corner

The baby animal petting corner at the country show is going to be a **6 metre by 8 metre rectangle**.

Within that space:

- the chicks need a 2 metre by 2 metre square enclosure
- the kid goats need 12 square metres
- the lambs need 9 square metres
- the calves need 4 times as much space as the chicks.

The Problem

The baby animal petting corner organiser is getting stressed and needs help to work out:

- a plan for each enclosure
- how much outside fencing is needed, allowing for a 1 metre entrance and 1 metre exit
- how much internal fencing is needed
- how many 1 metre square pavers are needed to make a path through the baby animal petting corner.

The Challenge

Use the grid below to make up a plan for the organiser. Show how much fencing and how many pavers are needed, and where the entrance and exit gates should go.

But be quick – the country show opens next week.

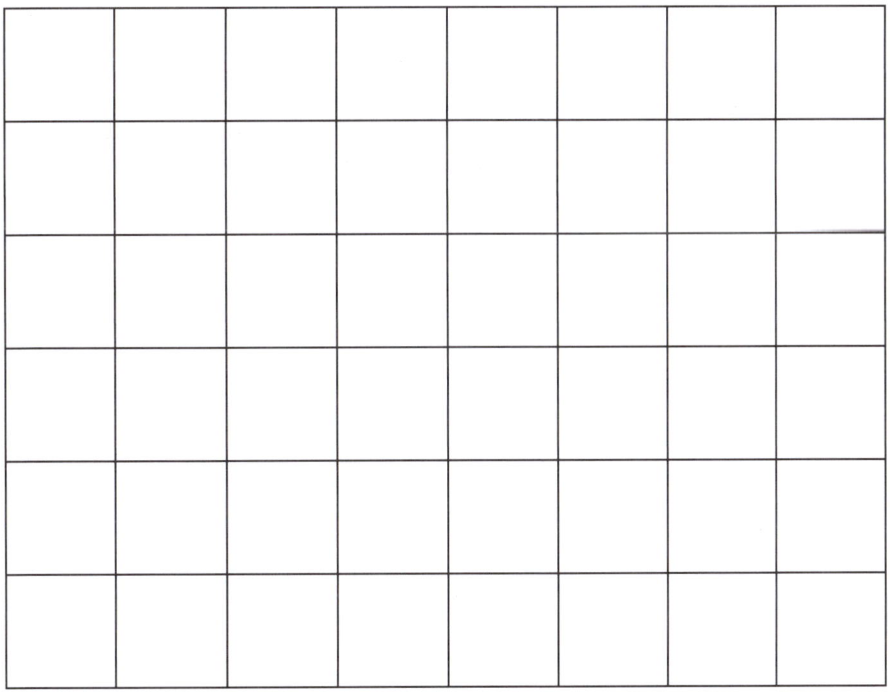

The Solution

The organiser will need:

- ______ metres of external fencing
- ______ metres of internal fencing
- ______ pavers for the path.

AC9M5N06

Preparing for Calculator Errors

Calculators do not make errors.

It is us, the calculator operators, who might make an operator error. That is why it is always important to identify the possible answer range before you reach for the calculator.

For example, to find 365 × 6 we could:

- **Round** 365 **down** to 300 to have a **low estimate** of 300 × 6 = 1800.
- **Round** 365 **up** to 400 to have a **high estimate** of 400 × 6 = 2400.
- **Adjust** these estimates, knowing that 365 is closer to 400 than to 300 to make a closer estimate of, say, 2200.

Now enter the multiplication on the calculator: 3 6 5 × 6 = 2190

For these multiplications, make a low estimate and a high estimate.
Then adjust the estimates to make a closer estimate.
Last of all, do the multiplication on your calculator and enter the calculator result.

	Multiplication	Round down: low estimate	Round up: high estimate	Adjust	Calculator Result
1	4 5 9 × 7 =	400 × 7 = ______	500 × 7 = ______		
2	2 5 9 × 8 =				
3	3 7 2 × 4 =				
4	5 3 4 × 8 =				
5	1 4 7 6 × 7 =				

SCORE /5 0-2 3-4 5

Quick Check

It is good to keep practising your tables so, for these questions, set a timer to see how long you take. Tick the questions that you instantly recalled and practise those that you didn't tick.

Total Time Taken = ______

1. 6 × 6 =
2. 7 × 8 =
3. 4 × 9 =
4. 5 × 7 =
5. 6 × 8 =
6. 8 × 8 =
7. 9 × 9 =
8. 7 × 9 =
9. 9 × 6 =
10. 3 × 7 =
11. 7 × 7 =
12. 6 × 7 =
13. 9 × 5 =
14. 12 × 3 =

SCORE /14 0-6 7-12 13-14

TERM 3

TARGETING MATHS HOMEWORK: YEAR 5 © PASCAL PRESS ISBN: 9781925726589

Number & Algebra

AC9M5N06, AC9M5N08

Multiplying Money Amounts

We can also use rounding when multiplying money amounts.
For example, to find the cost of 6 items at $49·50 each:

- **Round** $49·50 up to $50 by adding 50c and then multiply by 6.
  We know that 6 × $50 = $300.
- **Adjust** by subtracting 6 lots of 50c which is $3 to find the answer.
  $300 – $3 = $297

Round and adjust to complete these money multiplications.

① 8 at $19·95 each

8 × $20 = __________

8 × 5c = __________

_______ – _______ = _______

② 4 at $29·75 each

___ × _____ = _______

___ × _____ = _______

_______ – _______ = _______

③ 12 at $9·90 each

___ × _____ = _______

___ × _____ = _______

_______ – _______ = _______

④ 7 at $30·20 each

___ × _____ = _______

___ × _____ = _______

_______ + _______ = _______

⑤ 6 at $120·10 each

___ × _____ = _______

___ × _____ = _______

_______ + _______ = _______

⑥ 8 at $45·55 each

___ × _____ = _______

___ × _____ = _______

_______ + _______ = _______

SCORE /6 0-2 3-4 5-6

Statistics & Probability

AC9M5P02

Changing Chance

I put ten counters labelled 1, 2, 3, 4, 5, 6, 7, 8, 9 and 10 into this bag. Then I gave the bag a shake.

① What is the chance that (3) is the counter that I pull out of the bag?

____

② I put that counter back in the bag and removed these counters: (1) (6) (8)

What is the chance of pulling out a 3 now?

____

③ I removed three more counters from the bag: (2) (5) (9)

What is the chance of pulling out a 3 now?

____

④ Next, I take these counters out of the bag: (4) (7) (10)

What is the chance of pulling out a 3 now?

____

SCORE /4 0-1 2-3 4

AC9M5SP02

Where Is the Treasure?

The treasure is hidden on this grid. I have made up some clues to help you find it. The clues will tell you which points on the grid can be eliminated from your search. For example, my first clue is: **The treasure is not at (0, 0).**
So I have eliminated that position for you.

Follow Will's clues to find the treasure by eliminating points.

1. The first number in the coordinate pair is 2 or more.
2. You would have to travel north or east of (3, 3) to find the treasure.
3. The treasure is closer to the east side of the grid than it is to the west side of the grid.
4. Both numbers in the coordinate pair are the same.
5. Both numbers in the coordinate pair are factors of 10.
6. Where is the treasure?

   At (__ , __)

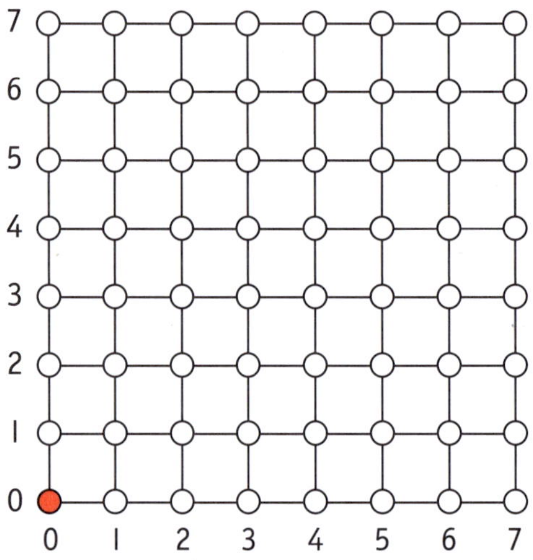

SCORE /6 0-2 3-4 5-6

NAPLAN*-style Questions

1. Coach bought 8 new footballs at $27·95 each for team practice.
   Which of these is closest to the amount that Coach paid?

   ◯ $35·95 ◯ $224 ◯ $160 ◯ $250

2. Asha labelled three counters with 1, three counters with 2 and three counters with 3. Then she put all nine counters in a bag.

   What is the chance that she will pull out a counter labelled 2?

   ◯ $\frac{2}{3}$ ◯ $\frac{1}{9}$ ◯ $\frac{1}{2}$ ◯ $\frac{1}{3}$

3. Join these coordinate pairs on the grid in the order listed.

   (1, 3), (1, 4), (1, 5), (2, 5), (3, 5), (3, 4), (3, 3) (2, 3), (1, 3)

   What is the coordinate pair that lies in the middle of the shape that you have made?

   (___, ___)

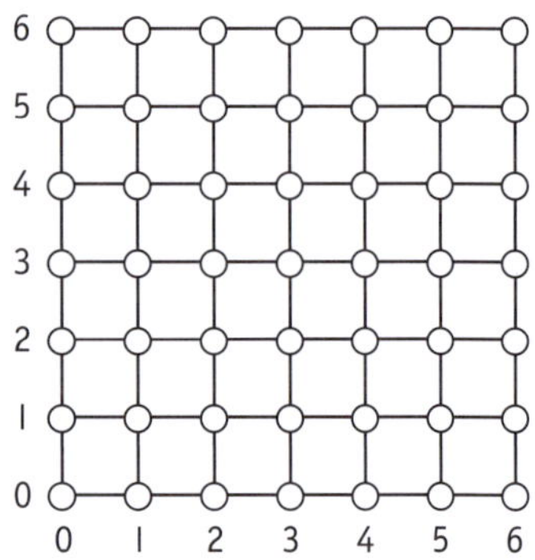

SCORE /3 0-1 2 3

TARGETING MATHS HOMEWORK: YEAR 5 © PASCAL PRESS ISBN: 9781925726589

Problem of the Week

AC9M5N06

Broken Calculator Keys

Oh no! My baby cousin has been playing with my calculator and has broken the **6** key. I have to use my calculator to finish my homework so I need to invent a work-around. Last time it was the **4** key that broke and I had to use a work-around to find **358 × 4**. I keyed in:

[3] [5] [8] [×] [5] [−] [3] [5] [8] [=]

and there was the answer in the display: 1432.

Write the key sequences to find these multiplications using a work-around for the broken 6 key. Ravi has started the first one for you.

	Multiplication	Key Sequence
①	345 × 6	[ ] [ ] [ ] [×] [1] [2] [÷] [ ] [=]
②	768 × 5	[ ] [ ] [ ] [ ] [ ] [ ] [ ] [ ] [=]
③	556 × 6	[ ] [ ] [ ] [ ] [ ] [ ] [ ] [ ] [ ] [ ] [ ] [=]
④	6666 × 9	[ ] [ ] [ ] [ ] [ ] [ ] [ ] [ ] [ ] [ ] [ ] [=]

What if more keys are broken?

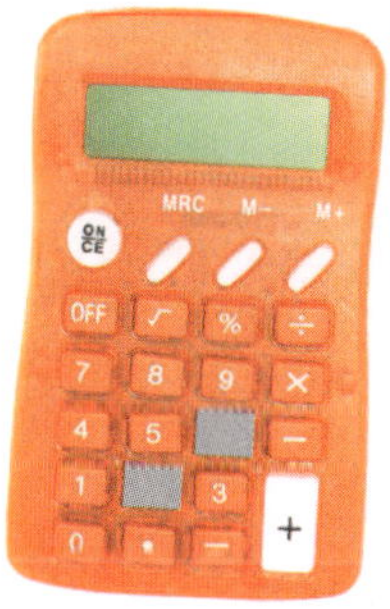

⑤ Find a work-around key sequence for **246 × 6** when the **2** key and the **6** key are broken.

⑥ Find a different work-around key sequence for **246 × 6** when the **2** key and the **6** key are broken.

⑦ Find a work-around key sequence for **762 × 27** when the **2** key and the **6** key are broken.

AC9M5N07

Division with Decimal Remainders

Sometimes the answer to a division question will have a decimal remainder.

If you use a calculator to find 6 ÷ 4, the key sequence will be:

[6] [÷] [4] [=] [1·5]

We often want the remainder for measurement or money divisions to be in decimal form.

For example, $6 shared between 4 people is $1·5 each, and we know that $1·5 is actually $1 and 50c.

6 metres of ribbon cut into 4 equal lengths is 1·5 metres for each piece and that is 1 metre and 50 cm.

Use your calculator to work out these divisions.
Write the answers in dollars and cents or in metres and centimetres.

1. $8 shared equally between 5 people.
   = _____ · _______ or $____ and ____ cents
2. 9 metres of fencing made into 4 fence panels
   = _____ · _______ or ____ m and ____ cm
3. $13 shared equally between 4 people
   = _____ · _______ or ____ and ____ cents
4. 5 metres of ribbon made into 4 bows
   = _____ · _______ or ____ m and ____ cm
5. $7 shared equally between two friends
   = _____ · _______ or $____ and ____ cents
6. $13 for 5 packets of balloons is how much for each packet?
   = _____ · _______ or $____ and ____ cents
7. If it takes 19 metres of tape to cordon off 5 equal sides of an area, how long is each side?
   = _____ · _______ or ____ m and ____ cm
8. If it is $21 for 6 smoothies, how much does each smoothie cost?
   = _____ · _______ or $____ and ____ cents

SCORE /8 (0-3) (4-6) (7-8)

Quick Check – Speedy Division Facts

For these questions, set a timer to see how long you take.
Tick the questions that you instantly recalled and practise those that you didn't tick.
Remember that you can use multiplication facts to work out division.

1. 15 ÷ 3 = ☐
2. 27 ÷ 9 = ☐
3. 64 ÷ 8 = ☐
4. 56 ÷ 7 = ☐
5. 81 ÷ 9 = ☐
6. 36 ÷ 4 = ☐
7. 72 ÷ 8 = ☐
8. 45 ÷ 5 = ☐
9. 49 ÷ 7 = ☐
10. 36 ÷ 6 = ☐
11. 28 ÷ 7 = ☐
12. 63 ÷ 9 = ☐
13. 27 ÷ 3 = ☐
14. 32 ÷ 4 = ☐
15. 60 ÷ 5 = ☐

SCORE /15 (0-7) (8-13) (14-15)

TERM 3

TARGETING MATHS HOMEWORK: YEAR 5 © PASCAL PRESS ISBN: 9781925726589

Number & Algebra

AC9M5N07, AC9M5N08

Number Splitting for Division

I find it really helpful to use my multiplication facts and number splitting to do division when large numbers are involved.

For example, to find 128 ÷ 4, I start with splitting the number 128 into two parts that I know to be multiples of 4.

Then I divide both parts by 4 and add the results to get the answer.

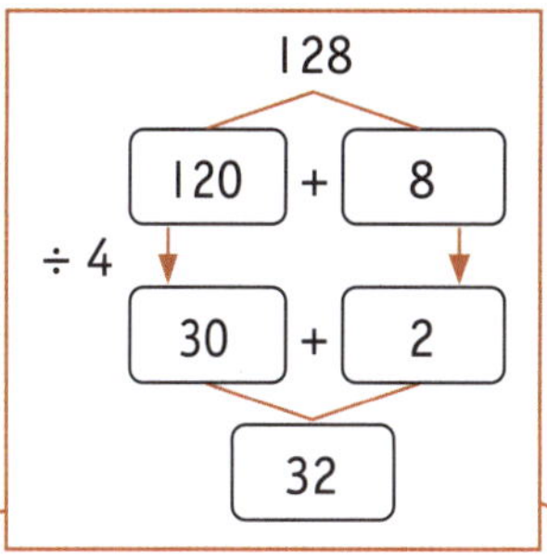

Use number splitting for these divisions.

1. 246 ÷ 6

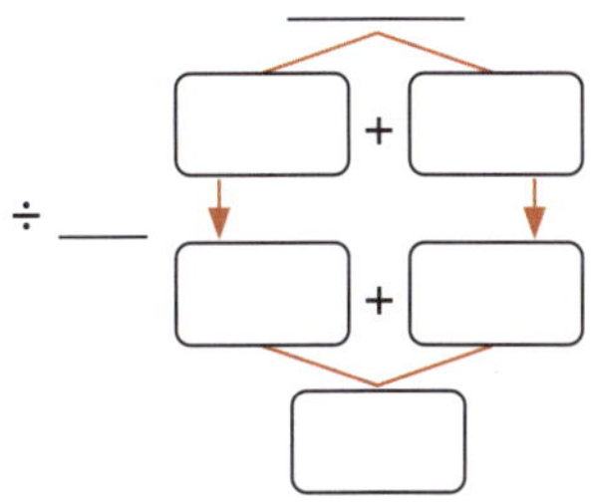

3. 642 ÷ 6

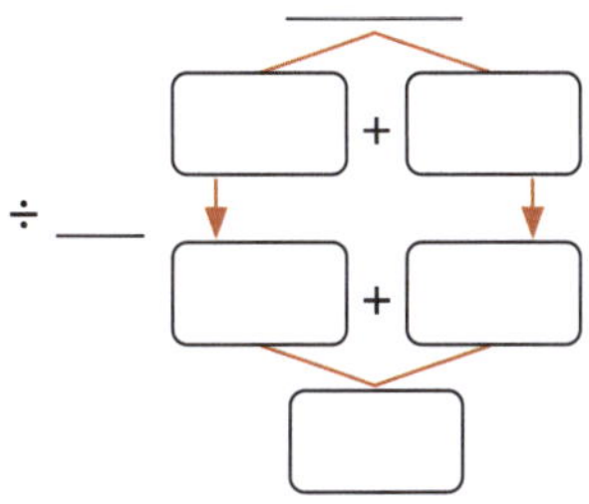

5. 492 ÷ 4

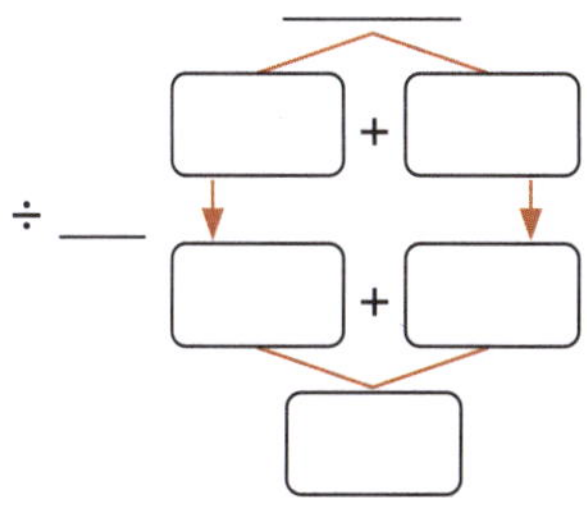

2. 357 ÷ 7

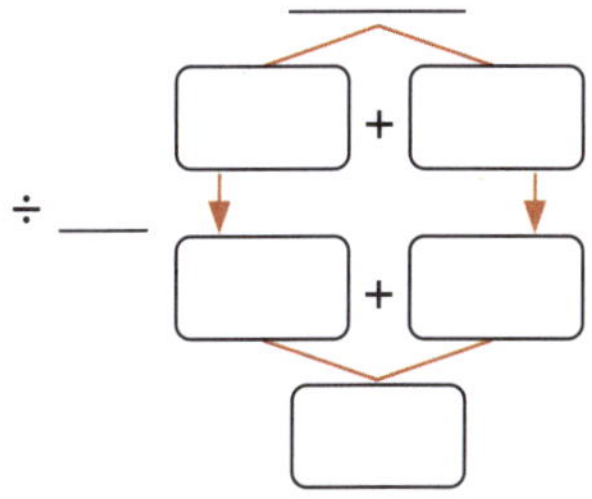

4. 96 ÷ 4

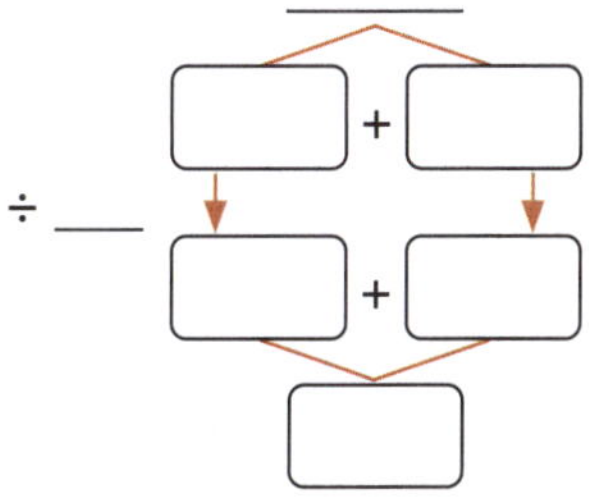

6. 572 ÷ 11

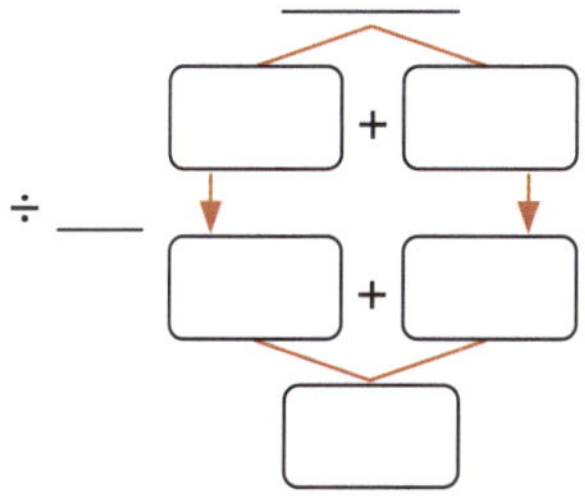

SCORE /6 0-2 3-4 5-6

Statistics & Probability

AC9M5P02

Which Colour?

Jade made this spinner in which each of the blue segments is $\frac{1}{6}$ of the whole.

1. Which colour has the best chance of being spun?

   ◯ Yellow ◯ Red ◯ Blue

2. Why did you choose that answer?

   ____

SCORE /2 0 1 2

Measurement & Space

AC9M5M02

Mixed Unit Conversions

When we write 1·5 kg, we know that means a **mass** of 1 kilogram and 500 grams.
Also, 1 kg is equal to 1000 grams and so 1·5 kg is equivalent to 1500 grams.
Similar conversions can be made with **length** where 1·5 m means 1 metre and 50 centimetres, and with **capacity** where 1·5 L means 1 litre and 500 millilitres.

Rename these measurements using mixed units.

1. 1·25 L = 1 litre and 250 millilitres or ________ mL
2. 2·36 kg = ______________________________ or ________ g
3. 3·05 L = ______________________________ or ________ mL
4. 1·065 kg = ______________________________ or ________ g
5. 1·7 cm = ______________________________ or ________ mm

Rename these measurements using mixed units and as a single unit.

6. 107 cm = 1 m and ___ cm or 1·07 m
7. 1250 g = __________________ or ________ kg
8. 2705 mL = __________________ or ________ L
9. 678 cm = __________________ or ________ m
10. 3005 g = __________________ or ________ kg

SCORE /10 0-4 5-8 9-10

NAPLAN*-style Questions

1. The grocer had 62 oranges which he put into nets of 15 oranges.
   How many oranges were left over?

   ☐

2. How many grams of flour are on the scales?

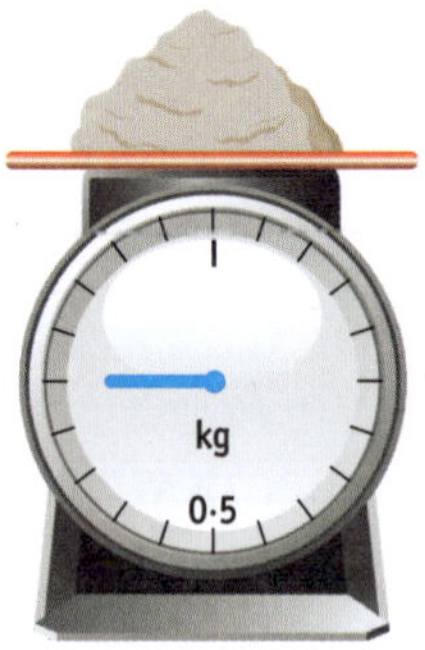

   - ◯ 125 g
   - ◯ 1500 g
   - ◯ 1250 g
   - ◯ 750 g

3. Which spinner has the best chance of spinning red?

   ◯ ◯ ◯

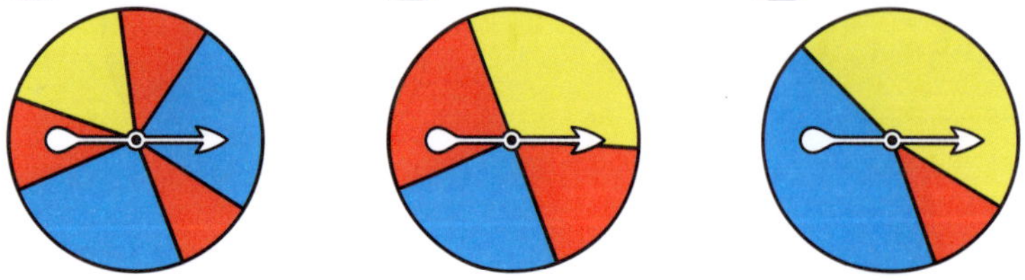

SCORE /3 0-1 2 3

TARGETING MATHS HOMEWORK: YEAR 5 © PASCAL PRESS ISBN: 9781925726589

Family Pet

Our family have decided to get a pet, but I can't decide what that pet should be.
I have this list of possible pets:

- cat
- goldfish
- rabbit
- guinea pig
- parrot
- dog.

I am going to try this experiment. I will write the types of pets on slips of paper.
I will make 6 slips for the pet I think will be the most popular, 5 slips for the pet that would be very popular, and so on, down to only 1 slip for the least popular pet.

Ella would like your help.

Please put the pets in order of popularity in the list below.
Then make 6 slips for the most popular down to 1 slip for the least popular.

Rating	Most popular	Very popular	Popular	Quite popular	A bit popular	Least popular
Type of pet						

Try the Experiment

- Put your slips of paper into a bag and give the bag a good shake.
- Take 10 slips of paper out of the bag.
  Record the number of slips that you have for each pet in the table below.
- Repeat the experiment until you have 4 sets of data.
  Then suggest to Ella which three pets the family should choose from.

	Cat	Goldfish	Rabbit	Guinea pig	Parrot	Dog
Experiment 1						
Experiment 2						
Experiment 3						
Experiment 4						
Totals						

Your results

1. What are the three pets that your data suggest that Ella's family should choose from?

   Pet 1: ____________________

   Pet 2: ____________________

   Pet 3: ____________________

2. How close was your rating of the pets' popularity to the results of the experiments?

   ____

Division with Remainders

For the division problems with remainders below, you will need to decide which kind of remainder is needed.

Example 1

27 apples are to be divided equally between 4 trays.
The answer will be 6 in each tray with a remainder of 3 apples.

Example 2

13 metres of string divided into 4 equal lengths could be 3 metres each with a remainder of 1 metre.

But that remainder could be further divided into 4 parts, so the answer should be $1\frac{1}{4}$ metres or 1·25 metres.

For each division, write the answer and the remainder that makes the most sense. Note that there might be two possible ways of recording the remainder.

1. 26 oranges to be divided equally into 4 nets is

   ____

2. Will cut up 5 blocks of cheese to share equally between 4 trays of mac and cheese, ready to be frozen.

   How much cheese did Will use for each tray of mac and cheese?

   ____

3. 4 kilograms of flour shared between four 800-gram mixing bowls is

   ____

4. $15·05 to be shared fairly between 3 friends is

   ____

5. 1 metre of ribbon cut into 30 cm pieces is

   ____

SCORE /5 (0-2) (3-4) (5)

Quick Check

Write the answer for each division as a number and a remainder.

1. 27 ÷ 7 = ________
2. 49 ÷ 6 = ________
3. 58 ÷ 7 = ________
4. 67 ÷ 8 = ________
5. 60 ÷ 9 = ________

Write the answer for each measurement division as a decimal and as a fraction.

6. $2\frac{1}{2}$ kg ÷ 2 = ________ or ________
7. 5 m ÷ 4 = ________ or ________
8. 7 m ÷ 4 = ________ or ________
9. 81 cm ÷ 4 = ________ or ________
10. 5 L ÷ 4 = ________ or ________

SCORE /10 (0-4) (5-8) (9-10)

TARGETING MATHS HOMEWORK: YEAR 5 © PASCAL PRESS ISBN: 9781925726589

Number & Algebra

UNIT 22

AC9M5A01

Using Number Splitting to Make Estimates of Divisions

To make an estimate of a division of a large number, I like to use multiplication facts to help me make a close estimate of the answer. Then I can check with my calculator.

To make an estimate of **362 ÷ 5**, I use the multiplication 35 ÷ 5 = 7 to make a number split. Then I make an estimate with the rest.

My estimate for 362 ÷ 5 is 72.

Then I check with the calculator: 3 6 2 ÷ 5 = 72·4

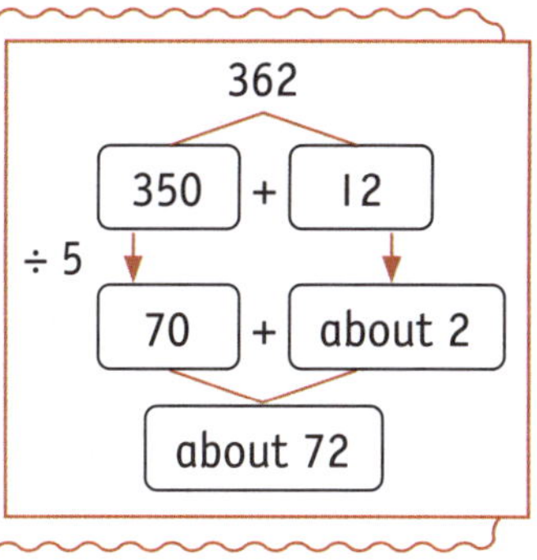

Complete these divisions using number splitting to make your estimates and then checking with your calculator.

1. 498 ÷ 8 ______
   [ ] + [ ]
   ÷ __
   [ ] + about ___
   about ____
   Calculator answer = ______

2. 364 ÷ 7 ______
   [ ] + [ ]
   ÷ __
   [ ] + about ___
   about ____
   Calculator answer = ______

3. 653 ÷ 5 ______
   [ ] + [ ]
   ÷ __
   [ ] + about ___
   about ____
   Calculator answer = ______

4. 513 ÷ 6 ______
   [ ] + [ ]
   ÷ __
   [ ] + about ___
   about ____
   Calculator answer = ______

5. 459 ÷ 6 ______
   [ ] + [ ]
   ÷ __
   [ ] + about ___
   about ____
   Calculator answer = ______

6. 784 ÷ 7 ______
   [ ] + [ ]
   ÷ __
   [ ] + about ___
   about ____
   Calculator answer = ______

SCORE /6 0-2 3-4 5-6

Statistics & Probability

AC9M5ST03

Interpreting a Dot Plot

This is a chart of the results of the 100 m race trials.

1. How many runners took part? ____
2. What was the difference between the fastest and slowest times? ______________
3. How long did only one runner take? ______________
4. What fraction of the runners took more than 17 seconds? ____

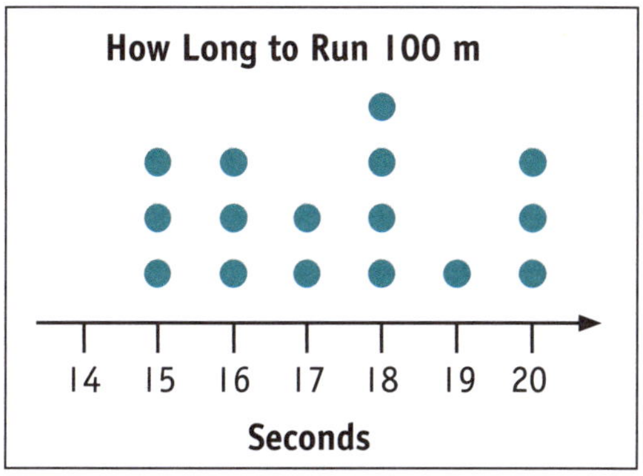

SCORE /4 0-1 2-3 4

AC9M5M04

Angles of a Triangle

I have marked the **internal angles** of a triangle. The internal angles always add to 180° – the shape of the triangles doesn't matter.

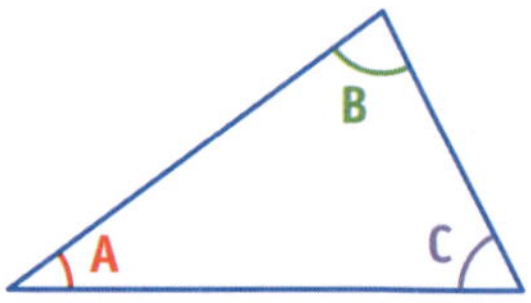

Try this and you will see why that is always true.

- Draw a triangle and cut it out.
- Draw a straight line on a piece of paper. Mark a point near the middle.
- Tear off each corner of the triangle. Arrange the pieces on your straight line so that each corner is on the point that you marked. The pieces shouldn't overlap but should touch another piece along one side.
- You will see that the pieces make a straight line angle of 180°.

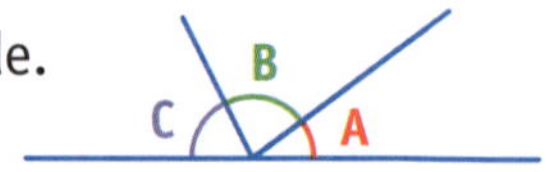

These triangles each have an unlabelled angle. What is the missing angle?

① ______°

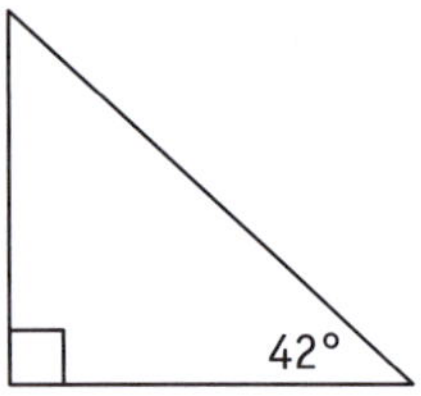

② ______°

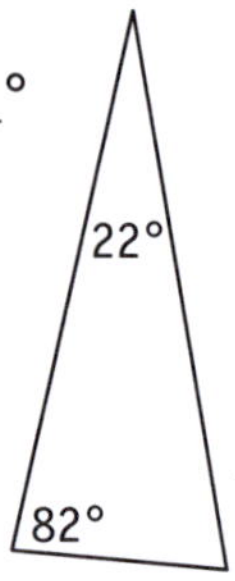

③ ______°

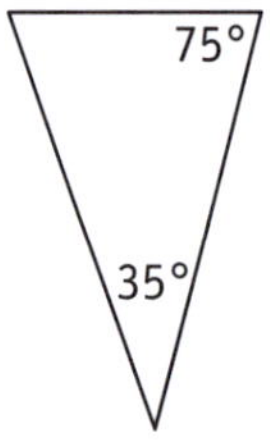

④ ______°

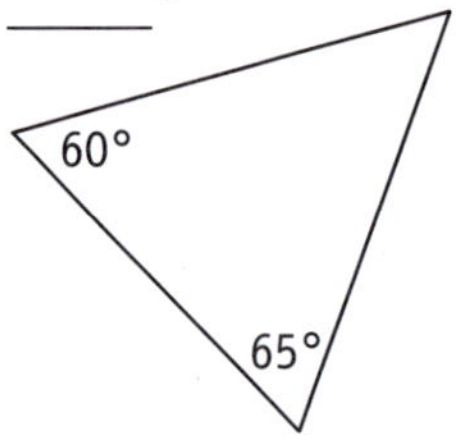

⑤ ______°

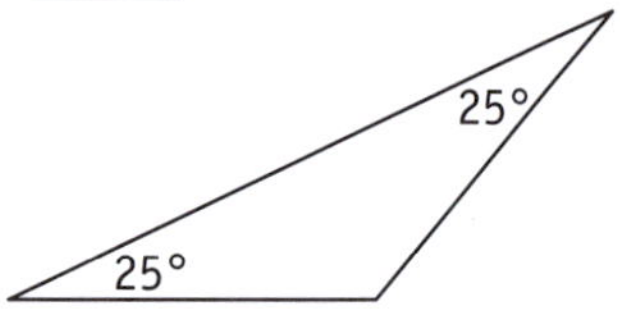

⑥ ______°

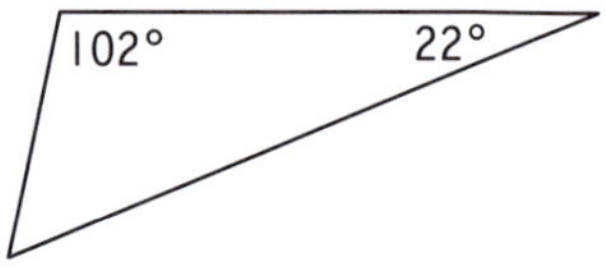

SCORE /6 0-2 3-4 5-6

NAPLAN*-style Questions

① Ravi bought 5 notebooks for $6. How much did each notebook cost?

- ◯ $11
- ◯ $30
- ◯ $1·50
- ◯ $1·20

② This dot plot shows the test scores out of 20.

How many students sat the test?

▢

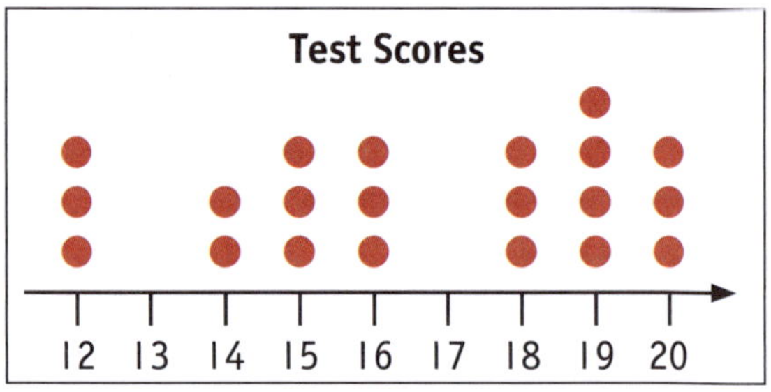

③ Which of the angles in this diagram is equal to 30°?

- ◯ A
- ◯ C
- ◯ B
- ◯ D

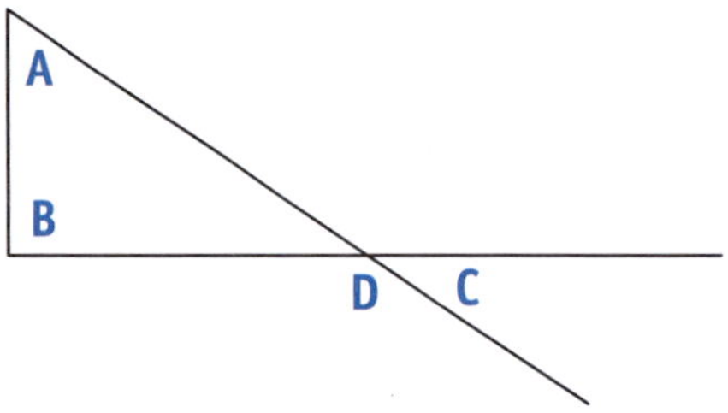

SCORE /3 0-1 2 3

TERM 3

TARGETING MATHS HOMEWORK: YEAR 5 © PASCAL PRESS ISBN: 9781925726589

Problem of the Week

AC9M5N07

Chef's Sushi Diner

Chef had a busy morning making 303 sushi rolls.

She made four different types of sushi rolls: avocado, prawn, chicken and beef.

"I will not tell you how many of each type I made, but I will give you some clues so you can work it out."

Clue 1 There are 30-something avocado rolls and the number could be divided by 2, but when it is divided by 3 there is a remainder of 1 and when it is divided by 8 there is a remainder of 2.

How many avocado sushi rolls did Chef make?

Show your thinking here.

Clue 2 There are 60-something prawn rolls and the number can be divided by 4 but when divided by 3 there is a remainder of 2.

Show your thinking here.

Clue 3 There are 120-something beef sushi rolls and the number can be divided by 2, 3, 6 and 9 with no remainder.

Show your thinking here.

How many chicken sushi rolls did Chef make?

Show your thinking here.

AC9M5N10

Exploring Multiples and Factors

You can use a diagram to explore the factors of a given number.

The diagram is called a factor tree, and it is a way to find factors of a number.

If we start with the number 18, we know that it can be made with 2 × 9 or 3 × 6.

For those products, the factor trees are:

18
2 × 9
3 × 3

18
3 × 6
2 × 3

The diagrams show that 2, 3, 6 and 9 are all proper factors of 18.

Complete the factor trees.

1. 12
   2 × ___
   ___ × ___

   12
   3 × ___
   ___ × ___

2. 28
   2 × ___
   ___ × ___

   28
   7 × ___
   ___ × ___

3. 24
   2 × ___
   ___ × ___
   ___ × ___

   24
   3 × ___
   ___ × ___
   ___ × ___

4. 36
   2 × ___
   ___ × ___
   ___ × ___

   36
   3 × ___
   ___ × ___
   ___ × ___

5. 42
   2 × ___
   ___ × ___

   42
   7 × ___
   ___ × ___

6. 54
   2 × ___
   ___ × ___
   ___ × ___

   54
   3 × ___
   ___ × ___
   ___ × ___

SCORE /6 0-2 3-4 5-6

Quick Check

What is the lowest common multiple (LCM)?

1. 3 and 4 _____
2. 2, 3 and 4 _____
3. 2, 4, and 5 _____
4. 2, 4, 6 and 8 _____
5. 3, 5, 9 and 15 _____

Write the factors.

6. 8 ____________________
7. 15 ____________________
8. 18 ____________________
9. 24 ____________________
10. 36 ____________________

SCORE /10 0-4 5-8 9-10

TARGETING MATHS HOMEWORK: YEAR 5 © PASCAL PRESS ISBN: 9781925726589

Number & Algebra

AC9M5A01, AC9M5A02

Unknowns in Division and Multiplication

In a division question such as 325 ÷ ▢ = 5, the numbers are not in an order that you can use on a calculator.

But you can rearrange it to 325 ÷ 5 = ▢ and then use your calculator:

[3] [2] [5] [÷] [5] [=] [65]

This shows that the answer to the division question is 325 ÷ 65 = 5.

But in a multiplication question such as 65 × ▢ = 325, the unknown is found by rearranging it to 325 ÷ 65 = ▢.

Then use your calculator:

[3] [2] [5] [÷] [6] [5] [=] [5]

This shows the answer is 65 × **5** = 325.

Rearrange these division and multiplication questions and then find the unknown. Use a calculator if you need to.

1. Rearrange 324 ÷ ▢ = 6
   as 324 ÷ ___ = _____
   and so 324 ÷ ___ = 6
2. Rearrange 456 ÷ ▢ = 76
   as _____ ÷ ___ = _____
   and so _____ ÷ ___ = 76
3. Rearrange 1872 ÷ ▢ = 24
   as ______ ÷ ___ = _____
   and so ______ ÷ ___ = _____
4. Rearrange 2379 ÷ ▢ = 39
   as ______ ÷ ___ = _____
   and so ______ ÷ ___ = ___
5. Rearrange 27 × ▢ = 324
   as ______ ÷ ___ = _____
   and so ______ × ___ = ___
6. Rearrange 46 × ▢ = 368
   as ______ ÷ ___ = _____
   and so ______ × ___ = ___
7. Rearrange 35 × ▢ = 665
   as ______ ÷ ___ = _____
   and so ______ × ___ = ___
8. Rearrange 47 × ▢ = 423
   as ______ ÷ ___ = _____
   and so ______ × ___ = ___

SCORE /8 0-3 4-6 7-8

TERM 3

Statistics & Probability

AC9M5ST01

Plotting Plant Growth

Snow peas and beans grow very quickly, as the data shows.

Day	Mon.	Tue.	Wed.	Thu.	Fri.
Snow Pea (cm)	5	7	8	9	10
Bean (cm)	4	6	8	11	12

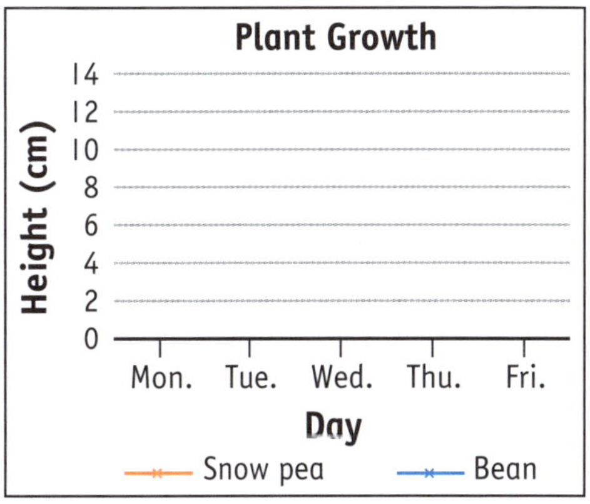

1. Use the two colours to create a line graph of the data.
2. On which day were the plants the same height?
   ____
3. How much more than the snow pea did the bean grow over the five days? ____ cm
4. Between which two days did the bean grow the most? ____________________
5. Between which two days did the snow pea grow the most? ____________________

SCORE /5 0-2 3-4 5

AC9M5M03

Using am, pm and 24-hour Clock Times

At five past three o'clock in the morning, these clocks would show:

12-hour digital time	24-hour digital time	Analogue time
3:05 am	03:05	

Use the time clues to work out what time should be showing on each clock.

Time Clues	12-hour digital time	24-hour digital time	Analogue time
① Ten minutes after five past three in the morning	: am pm	:	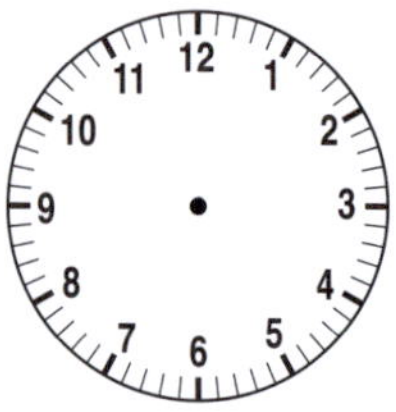
② Twenty minutes before five to four in the afternoon	: am pm	:	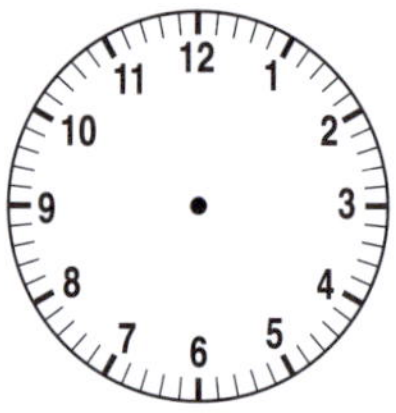
③ Twenty minutes later than ten to twelve at night	: am pm	:	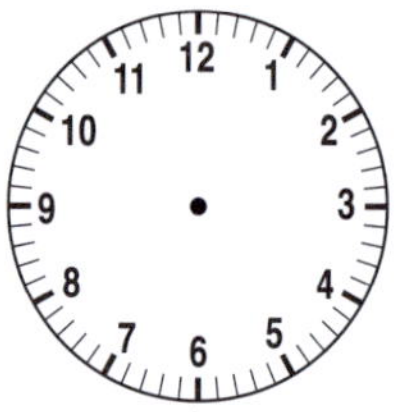
④ Twenty-two minutes later than twelve minutes after midday	: am pm	:	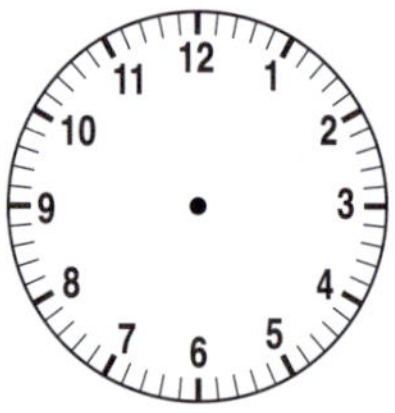

SCORE /12 0-5 6-10 11-12

TERM 3

NAPLAN*-style Questions

① Complete this multiplication: 19 × ____ = 95

② Which of these numbers is not a multiple of 7?

◯ 21 ◯ 14 ◯ 42 ◯ 12

③ This table shows the start and finish times of classic cartoons at the movie festival.

	Snow White	Dumbo	Pinocchio	Bambi
Start time	11:30 am	10:50 am	11:45 am	10:15 am
Finish time	12:55 pm	12:15 pm	1:15 pm	11:25 am

Which cartoon lasts the longest?

◯ Snow White ◯ Dumbo

◯ Pinocchio ◯ Bambi

SCORE /3 0-1 2 3

TARGETING MATHS HOMEWORK: YEAR 5 © PASCAL PRESS ISBN: 9781925726589

Problem of the Week

Time Pressure

Maria has to be at the International Airport two and a half hours before her flight to Italy leaves this morning at this time:

Maria thinks that she will:

- set the alarm for 

and allow:

- 20 minutes for last-minute packing and checking
- 10 minutes to ring her daughter and say goodbye
- 25 minutes to take the dog for a walk and leave her with the neighbours next door
- 20 minutes to book and wait for a taxi
- 35 minutes for the taxi ride to the airport.

Will she make it to the airport on time?

Hint: a timeline might help.

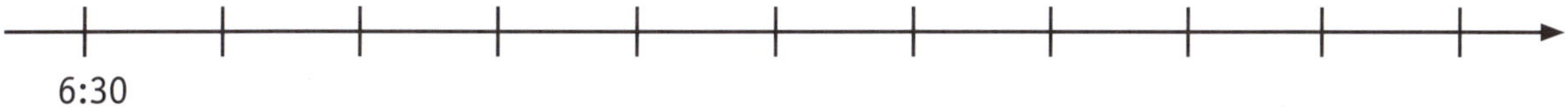

Guess what?

Maria suddenly realised that she had allowed no time for breakfast, or a shower or getting dressed.

She needs a rethink. What should she do?

Create a timetable that works.

Show the extra actions that Maria should include.

For what time should she set her alarm?

Alarm:

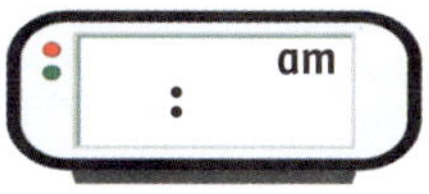

AC9M5N09

Financial Transactions

Problems that involve money changing hands often involve two steps to find the solution.

The problems below each have two steps for you to complete before you can work out the answer.

1. Jade bought three apples for $1·25 each and a pear for $1·50.
   She paid with a $10 note.
   How much change did she get?

   ______ × $ ______ = ______

   $______ + $______ = ______

   $10 – $______ = $______

2. Miro used a $20 note to pay for 6 pencils at 99c each and a notebook for $2·44.
   How much change did he get?

   ______ × $______ = ______

   $______ + $______ = ______

   $______ – $______ = $______

3. Ella gets $2·50 each day for walking the dog.
   She has $37·50 already saved and needs $65 for the T-shirt that she wants to buy.
   How many days until Ella can afford the T-shirt?

   $______ – $______ = ______

   $______ ÷ $______ = ______

   ___ days

4. Entrance to the zoo is $9·00 per person and there are 24 students signed up to go.
   The bus will be $240.
   How much will the excursion cost each student?

   ______ × $______ = ______

   $______ + $______ = $______

   $______ ÷ ______ = $______

5. There are actually 28 people going on the excursion to the zoo because 4 are supervising adults.
   The students have to pay for the adults' entrance to the zoo.
   How much will each student now need to pay for the excursion?

   ______ × $______ = ______

   $______ + $______ = $______

   $______ ÷ ______ = $______

SCORE /5 (0-2) (3-4) (5)

Quick Check

What is the change from $100 after spending these amounts?

1. $14·50. Change = $______
2. $18·75. Change = $______
3. $36·80. Change = $______
4. $74·25. Change = $______
5. $46·90. Change = $______

How much do these items cost?

6. 6 pencils at 35c. Cost $______
7. 1 muffin if 6 cost $24. Cost $______
8. 3 drinks at $2·85 each. Cost $______
9. 4 burgers at $8·35 each. Cost $______
10. 250 g of nuts at $36 per kg. Cost $______

SCORE /10 (0-4) (5-8) (9-10)

TARGETING MATHS HOMEWORK: YEAR 5 © PASCAL PRESS ISBN: 9781925726589

AC9M5N03

Fractions

Did you know that there are fractions that are smaller than one-tenth?

This rectangle has been cut into quarters.

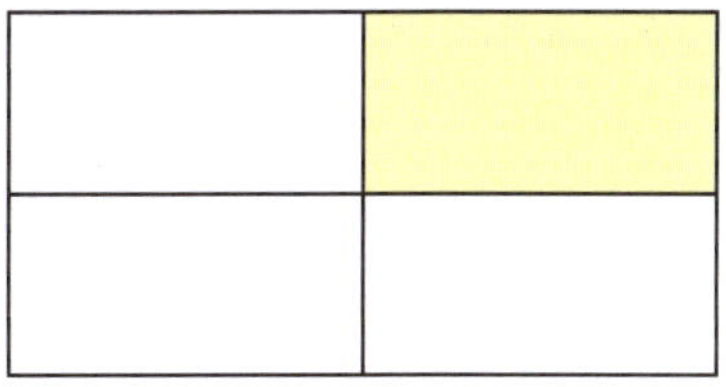

$\frac{1}{4}$

This rectangle has been cut into thirty-two parts.

$\frac{1}{32}$

We already know that $\frac{1}{2}$ is equivalent to $\frac{2}{4}$ and $\frac{4}{8}$.
The numerator and denominator of $\frac{1}{2}$ were multiplied by the same number to make the equivalent fractions (× 2 to make $\frac{2}{4}$ and × 4 to make $\frac{4}{8}$).

The same is true for other equivalent fractions.

$\frac{1}{3}$ is equivalent to $\frac{2}{6}$ after multiplying the numerator and denominator of $\frac{1}{3}$ by 2, and to $\frac{3}{9}$ after multiplying the numerator and denominator of $\frac{1}{3}$ by 3.

Continue these equivalent fraction strings.

1. $\frac{1}{1} = \frac{2}{2} = \frac{3}{3} = — = — = —$
2. $\frac{1}{2} = \frac{2}{4} = \frac{3}{6} = — = — = —$
3. $\frac{1}{3} = \frac{2}{6} = — = — = — = —$
4. $\frac{1}{5} = — = — = — = — = —$
5. $\frac{3}{4} = — = — = — = — = —$
6. $\frac{2}{3} = — = — = — = — = —$
7. $\frac{3}{5} = — = — = — = — = —$

SCORE /7 0-3 4-5 6-7

Statistics & Probability

AC9M5ST01

Choose a Colour

100 people were surveyed about the colour of their car.
20% said red, 15% said blue, half said yellow and the rest said green.
Only one of these charts does not match the survey data.

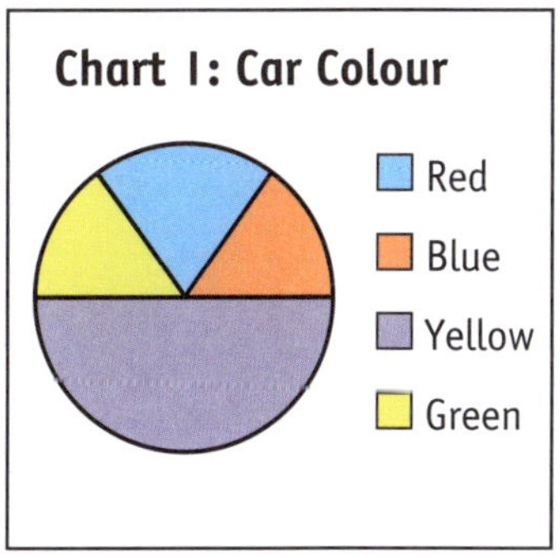

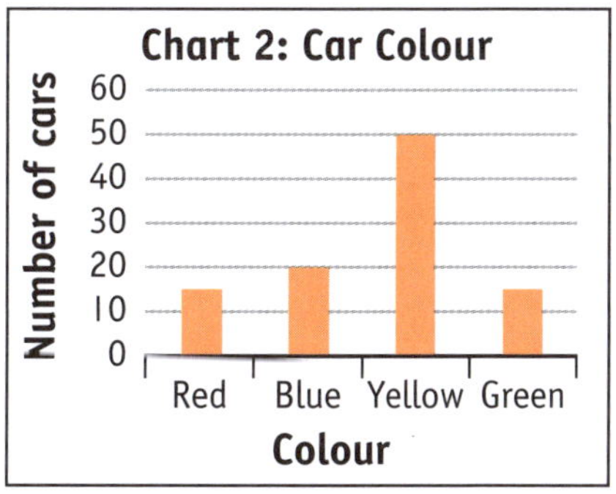

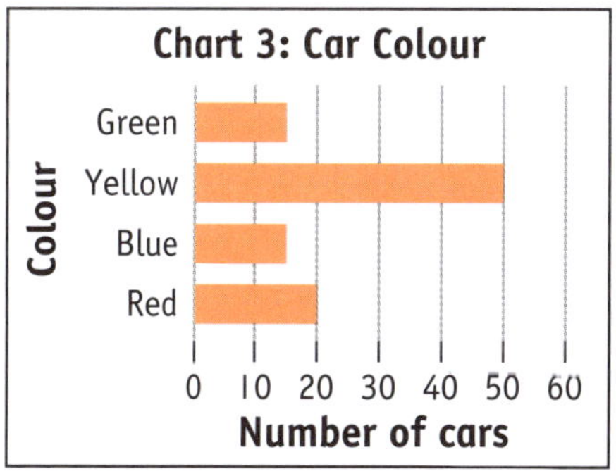

1. Which chart does not match the data? ______________
2. What is wrong with that chart? ______________________________

SCORE /2 0 1 2

AC9M5SP03

Translations, Reflections and Rotations

There are three changes you can make to the look of a shape without altering the size of it. The changes are called **transformations**.

For example, if you start with this shape drawn on a grid: 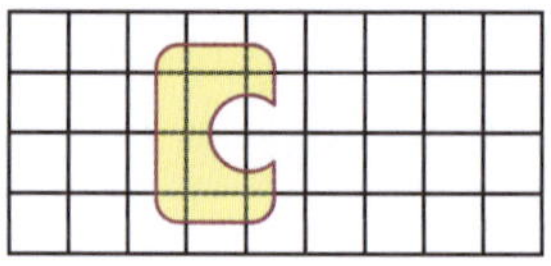

You can **slide** the shape: (Translation)

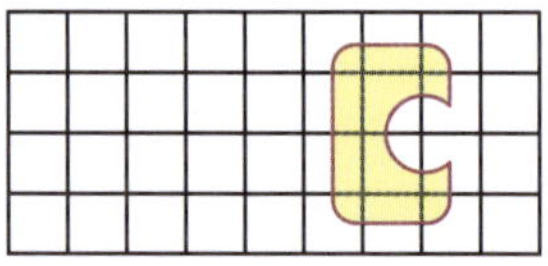

You can **flip** the shape: (Reflection)

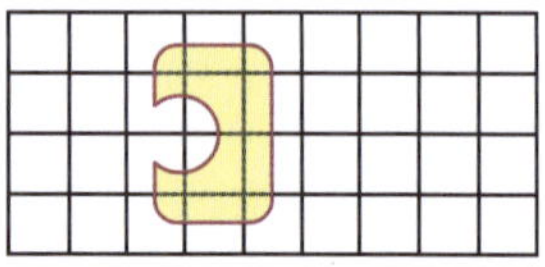

You can **turn** the shape: (Rotation)

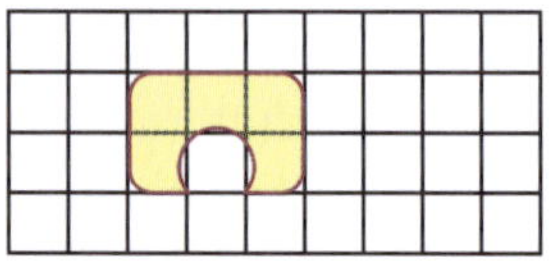

Leo wrote his name on a grid. Then he transformed each letter.

Before

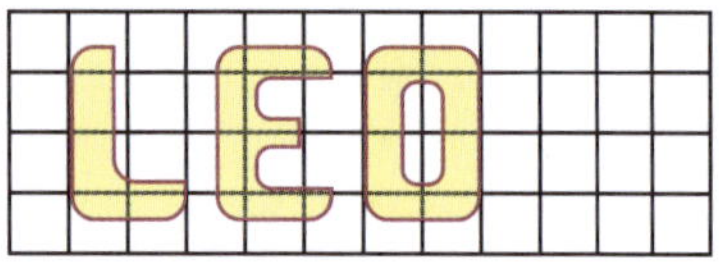

After

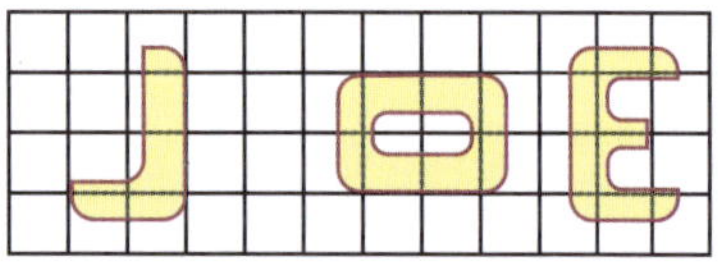

(1) What move did Leo use for the L? ______________

(2) What move did Leo use for the E? ______________

(3) What move did Leo use for the O? ______________

(4) **Write the first three letters of your name and then transform the letters.**

Before

After

List the moves you made for each letter.

(5) ______________

(6) ______________

(7) ______________

SCORE /7 0-3 4-5 6-7

NAPLAN*-style Questions

(1) What fraction is marked × on this fraction number line?

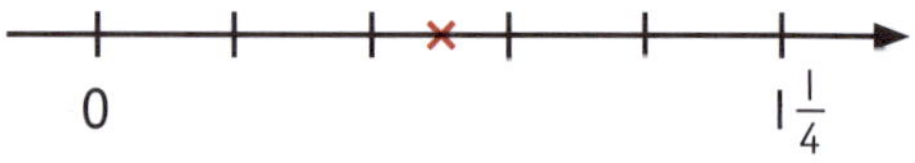

- $\frac{1}{2}$
- $\frac{1}{4}$
- $\frac{5}{8}$
- $\frac{6}{8}$

(2) Miro has saved $56 towards the scooter and helmet. How much more money does he need to save?

$69

$105

(3) Jade folded a piece of paper and then cut out the shape: What did she see when she opened up the cut-out piece?

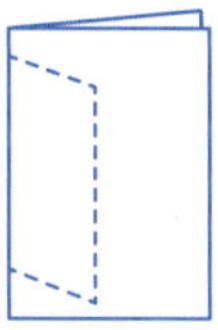

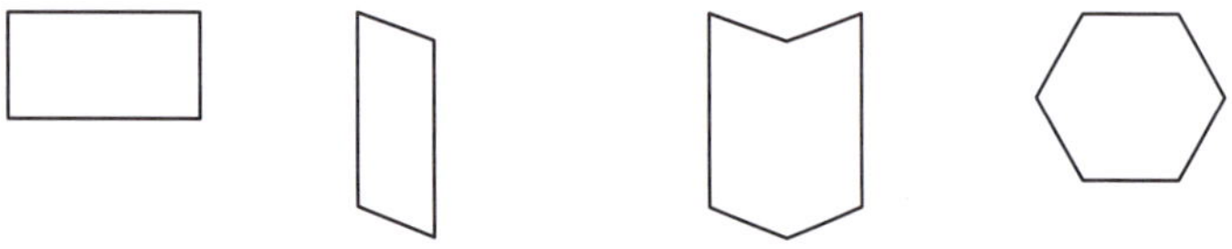

SCORE /3 0-1 2 3

TARGETING MATHS HOMEWORK: YEAR 5 © PASCAL PRESS ISBN: 9781925726589

Problem of the Week

AC9M5N08

Charlie the Builder

Charlie has lost the timesheets that his workers filled out at the end of the day, so he doesn't know how much to pay them.

He does know that he owes them $960 altogether and that they worked a total of 24 hours between them. Everyone is paid the same hourly rate.

The workers are not very happy with Charlie and so they want to make him work out how much each of them should be paid.

They gave Charlie these clues.

Yindi worked for a quarter of the hours.

Anh worked for 3 hours.

Liam worked for a third of the hours.

Archie does not remember how many hours he worked.

How much money does Charlie owe each of his workers?

Show all the steps in your thinking.

- Yindi should be paid $__________.
- Anh should be paid $__________.
- Liam should be paid $__________.
- Archie should be paid $__________.

TERM 3 REVIEW

Review: Unit 17 and Unit 18

AC9M5N01, AC9M5N02, AC9M5N03, AC9M5N04, AC9M5ST01

Convert these decimals into fractions.

(1) 0·8 =   (2) 0·89 =   (3) 0·089 =   (4) 1·89 =

Convert these fractions into decimals.

(5) $\frac{13}{100}$ = ______   (6) $\frac{8}{100}$ = ______   (7) $\frac{12}{1000}$ = ______   (8) $\frac{176}{100}$ = ______

Use ticks and crosses and the rules of divisibility to complete this table.

	The number	is divisible by						
		2	3	4	5	6	8	9
(9)	456							
(10)	6390							
(11)	5472							
(12)	4352							

What percentage of these batteries has not been used?

(13) 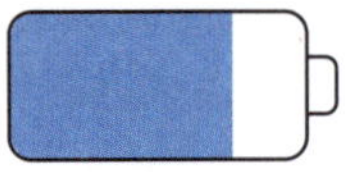

___% remains

(14) 

___% remains

(15) 

___% remains

(16) 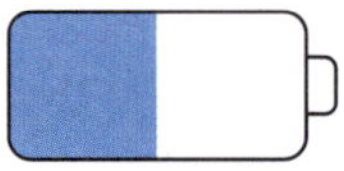

___% remains

This square is half of the whole shape.
What fractions of the whole shape do these diagrams represent?

(17) ____ 

(18) ____ 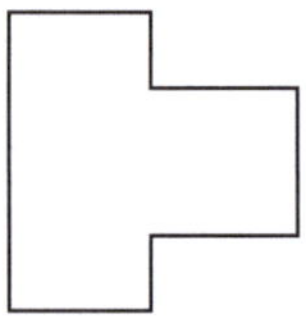

(19) ____ 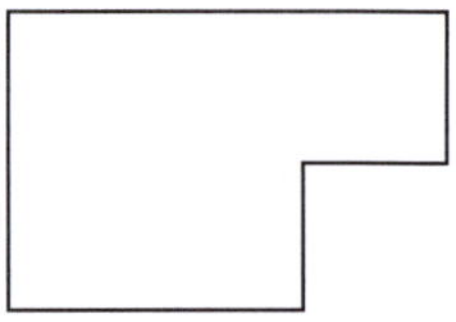

The chart shows the number of broken bones that were treated in six days.

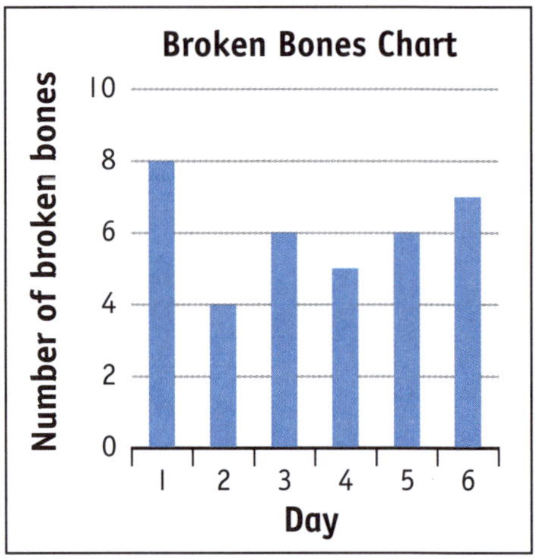

(20) How many broken bones were treated during the six days?

____

(21) What is the mode of the data? ____

(22) On which day were the most bones treated?

____

(23) How many more broken bones were treated in the first two days than in the second two days?

____

(24) Write a question of your own that the chart could answer.

____

SCORE /24 0-11 12-22 23-24

TERM 3

TARGETING MATHS HOMEWORK: YEAR 5 © PASCAL PRESS ISBN: 9781925726589

Review: Unit 19 and Unit 20

AC9M5N05, AC9M5N06, AC9M5N08, AC9M5M02, AC9M5SP02

Write these fractions in their equivalent form and then complete the additions and subtractions.

(1) $\frac{5}{8} + \frac{1}{4} = \frac{}{} + \frac{}{} = \frac{}{}$

(2) $\frac{3}{5} + \frac{1}{10} = \frac{}{} + \frac{}{} = \frac{}{}$

(3) $\frac{3}{6} + \frac{1}{2} = \frac{}{} + \frac{}{} = \frac{}{}$

(4) $\frac{3}{5} - \frac{1}{10} = \frac{}{} + \frac{}{} = \frac{}{}$

(5) $\frac{7}{8} - \frac{1}{4} = \frac{}{} + \frac{}{} = \frac{}{}$

(6) $\frac{5}{6} - \frac{1}{3} = \frac{}{} + \frac{}{} = \frac{}{}$

Write the perimeter and area of each of these shapes.

(7)

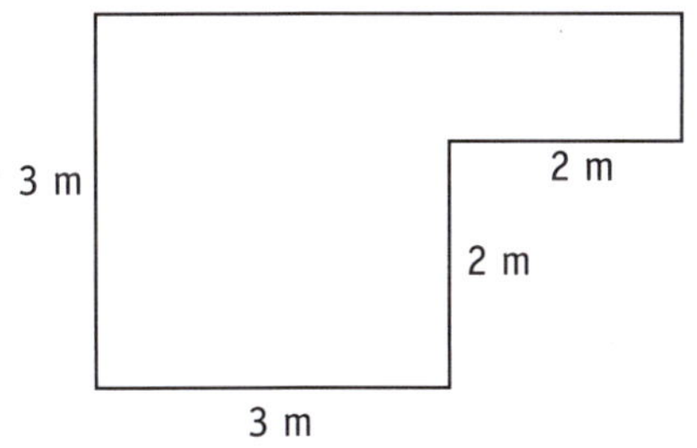

Perimeter = ________ Area = ________

(8)

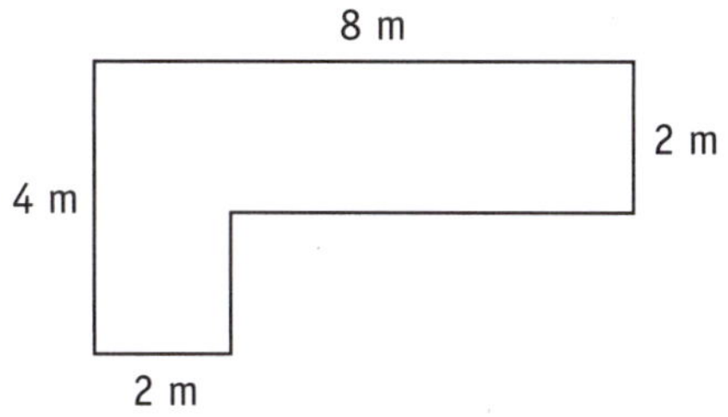

Perimeter = ________ Area = ________

Make a low and a high estimate for each of these money multiplications and then use your calculator to find the actual amount.

(9) $36·95 × 7 = ?

High estimate: ________________

Low estimate: ________________

Actual: ________________

(10) $3·75 × 6 = ?

High estimate: ________________

Low Estimate: ________________

Actual: ________________

Complete these money multiplications.

(11) 7 at $19·50

___ × _____ = _______

___ × _____ = _______

______ – ______ = _______

(12) 9 at $21·05

___ × _____ = _______

___ × _____ = _______

______ + ______ = _______

(13) Add the numbers to the axes of the grid.

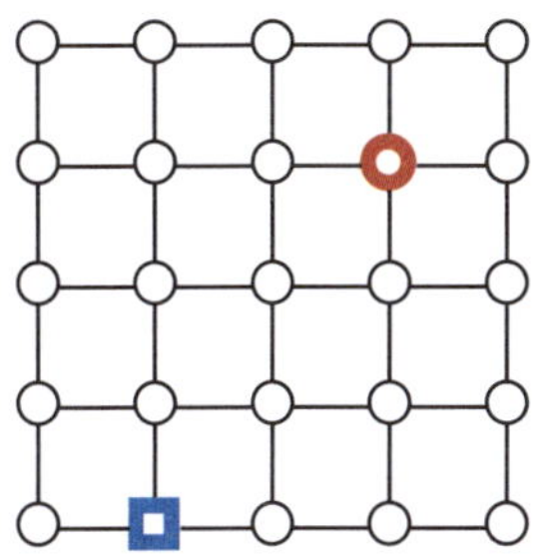

(14) What are the coordinates of ○? (___, ___)

(15) What are the coordinates of □? (___, ___)

SCORE /15 0-7 8-13 14-15

Review: Unit 21 and Unit 22

AC9M5N07, AC9M5N08, AC9M5N09, AC9M5M02, AC9M5M04, AC9M5A01

Use your calculator to work out these divisions.
Write the answers in dollars and cents or in metres and centimetres.

(1) $27 divided equally between 4 people = $______ or $___ and ____ cents each.

(2) 47 metres of fencing was made into a square garden fence.

Each side of the garden is ______ m or ___ m and ____ cm.

Use number splitting for these divisions.

(3) 175 ÷ 5 ______

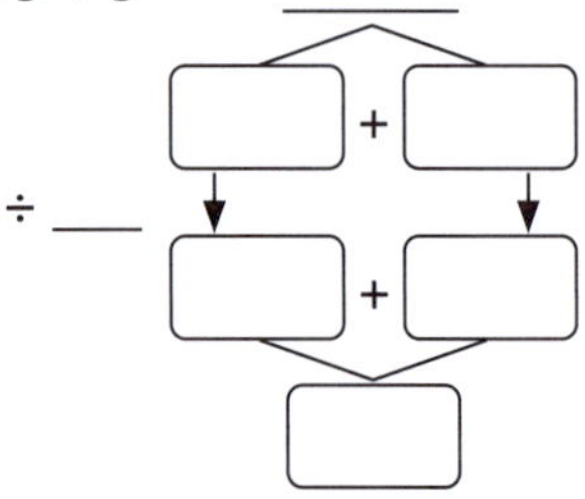

(5) 108 ÷ 9 ______

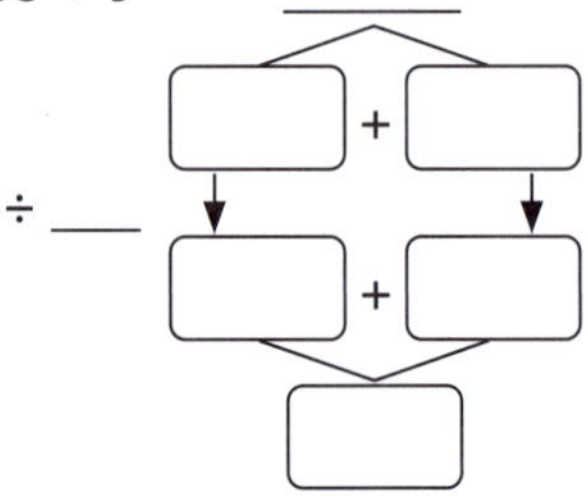

(4) 91 ÷ 7 ______

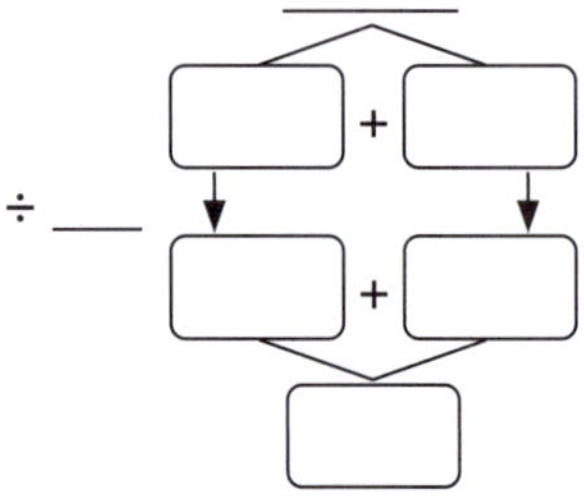

(6) 567 ÷ 9 ______

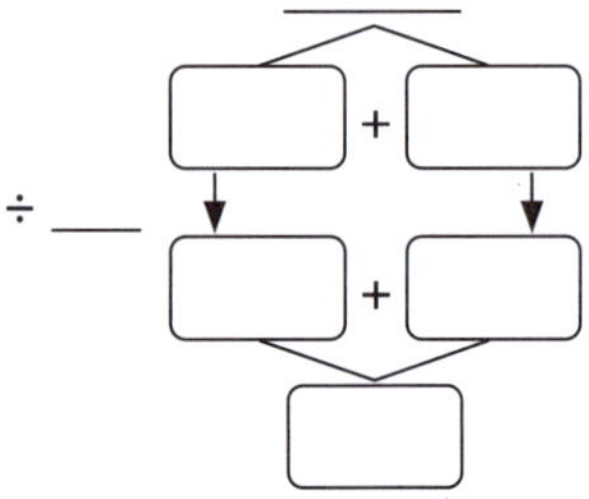

Rename these measurements using mixed units.

(7) 1·75 L = ___ L and ______ mL or ______ mL

(8) 3·75 kg = ___ kg and ______ g or ______ g

(9) 3·65 m = ___ m and ____ cm or ______ cm

Solve these division problems and decide which kind of remainder is needed.

(10) 38 apples to be put into 3 equal trays is ________________________________

(11) $17 to be shared between 4 people is ________________________________

Use number splitting to estimate the answers to these divisions.

(12) 552 ÷ 9 ______

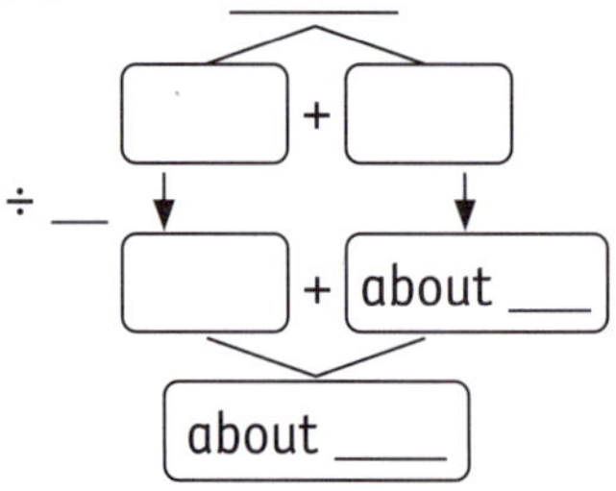

(13) 325 ÷ 3 ______

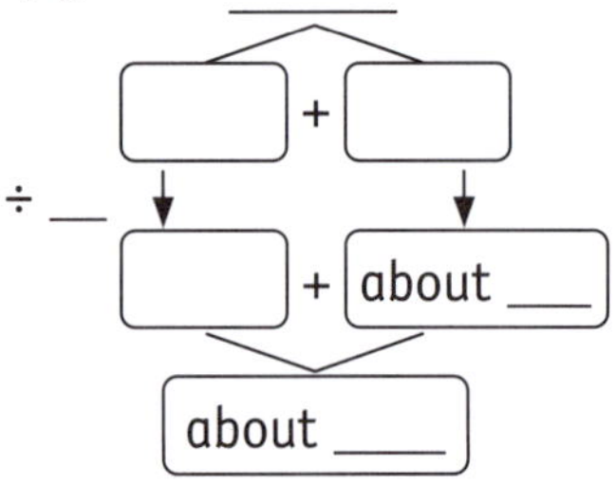

What are the missing angles in these triangles?

(14) ____

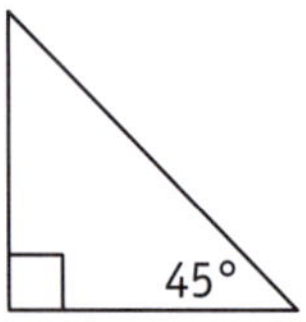

(15) ____

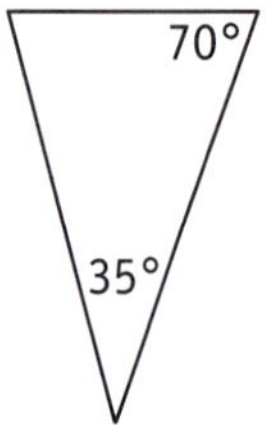

(16) ____

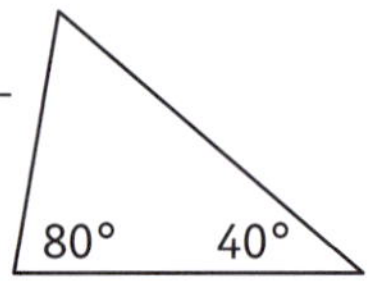

SCORE	/16	0-7	8-14	15-16

TARGETING MATHS HOMEWORK: YEAR 5 © PASCAL PRESS ISBN: 9781925726589

Review: Unit 23 and Unit 24

AC9M5N03, AC9M5N09, AC9M5N10, AC9M5M03, AC9M5SP03, AC9M5ST01

TERM 3 REVIEW

Complete the factor trees.

1. 32

___ × ___

___ × ___

___ × ___

___ × ___

2. 36

___ × ___

___ × ___

___ × ___

Use the graph at the right to answer the questions.

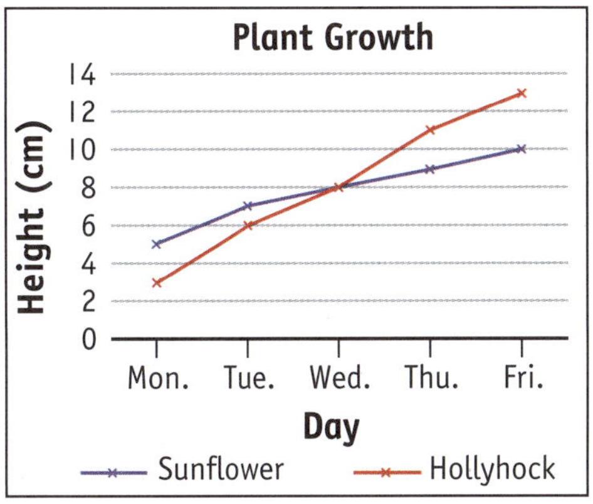

3. On which day were the sunflower and the hollyhock the same height?

____

4. Between Monday and Wednesday, how much more did the hollyhock grow than the sunflower?

____ cm

Show these times on the digital and analogue clocks.

	12-hour digital time	24-hour digital time	Analogue time
5. Quarter of an hour past 11:55 am	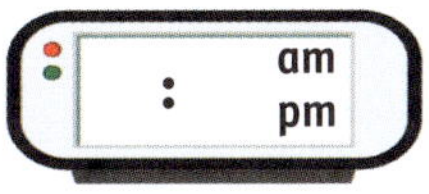		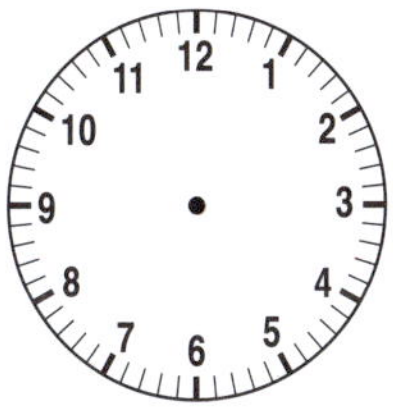
6. Twenty minutes before 12:10 am	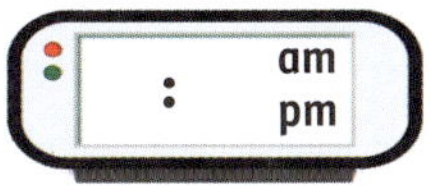		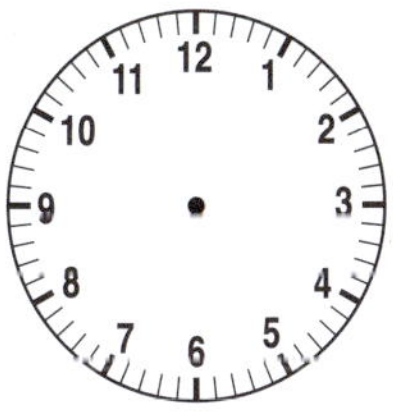

How much change from $50 after buying these items?

7. Two apples for $1·45 each

____

8. Six notebooks for $2·99 each

____

Continue these equivalent fraction strings.

9. $\frac{1}{5} = \frac{}{10} = \frac{3}{} = \frac{}{} = \frac{}{}$

10. $\frac{1}{6} = \frac{}{12} = \frac{3}{} = \frac{}{} = \frac{}{}$

11. Redraw the letters so that you translate A, reflect S and rotate H.

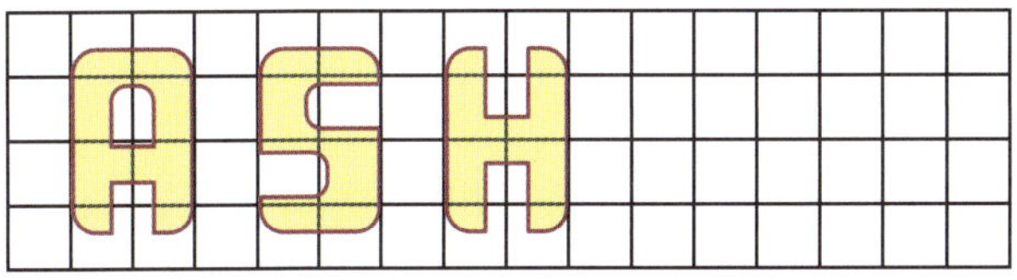

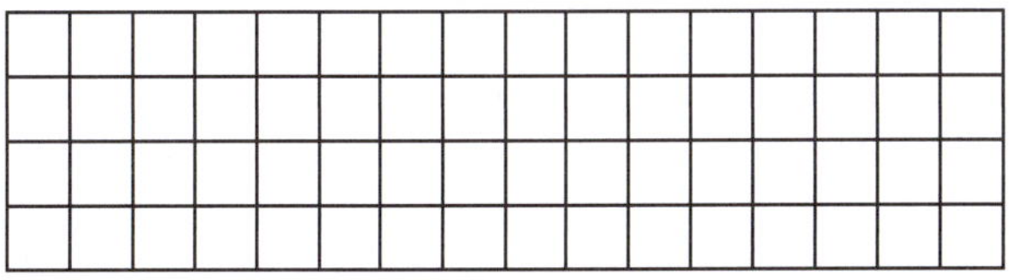

SCORE /11 0-5 6-9 10-11

TERM 3

Decimal Place Value

Order these fractions from smallest to largest.
Then write them in decimal form and complete the place value chart.

$\frac{375}{100}$ $\frac{456}{100}$ $\frac{37}{1000}$ $\frac{1456}{1000}$ $\frac{9}{1000}$

	Fraction	Decimal	Ones	·	Tenths	Hundredths	Thousandths
1				·			
2				·			
3				·			
4				·			
5				·			

Order the decimal numbers from smallest to largest. Then write them in expanded form.
For example, 0·49 in expanded form is 0 ones, 4 tenths and 9 hundredths.

3·76 3·079 2·685 2·608 3·009

6. ________ is ___ ones, ___ tenths, ___ hundredths and ___ thousandths.
7. ________ is ___ ones, ___ tenths, ___ hundredths and ___ thousandths.
8. ________ is ___ ones, ___ tenths, ___ hundredths and ___ thousandths.
9. ________ is ___ ones, ___ tenths, ___ hundredths and ___ thousandths.
10. ________ is ___ ones, ___ tenths, ___ hundredths and ___ thousandths.

SCORE /10 0-4 5-8 9-10

TERM 4

Quick Check

How many tenths in these numbers?

1. 3·65 has ____ tenths.
2. $\frac{35}{100}$ has ____ tenths.
3. $3\frac{15}{100}$ has ____ tenths.
4. 6·057 has ____ tenths.
5. $\frac{1567}{1000}$ has ____ tenths.

How many hundredths in these numbers?

6. 0·56 has ___ hundredths.
7. 3·05 has ___ hundredths.
8. 1·56 has ____ hundredths.
9. $\frac{156}{1000}$ has ____ hundredths.
10. $\frac{26}{10}$ has ____ hundredths.

SCORE /10 0-4 5-8 9-10

TARGETING MATHS HOMEWORK: YEAR 5 © PASCAL PRESS ISBN: 9781925726589

Number & Algebra

AC9M5N02

Factors and Multiples

Use the numbers in the strings to make factor and multiple sequences.

Here is an example that I made to start you off:

Reorder the numbers 2, 3, 4, 6, 8 and 12 with no repeats.

2 is a factor of 6.
6 is a multiple of 3.
3 is a factor of 12.
12 is a multiple of 4.
4 is a factor of 8.

Use these numbers to make factor–multiple sequences.

	Numbers to use	Factor	Multiple	Factor	Multiple	Factor	Multiple
1	2, 3, 5, 6, 15, 30						
2	2, 3, 6, 9, 12, 18						
3	2, 4, 8, 10, 12, 20						
4	2, 5, 10, 20, 50, 100						
5	6, 9, 18, 27, 36, 54						

SCORE /5 0-2 3-4 5

Statistics & Probability

AC9M5ST02

Line Graphs

Graphs can tell stories.
This line graph plots a car trip made by a family. Mum, Dad, Miro, Sam and their dog Ollie were all in the car.

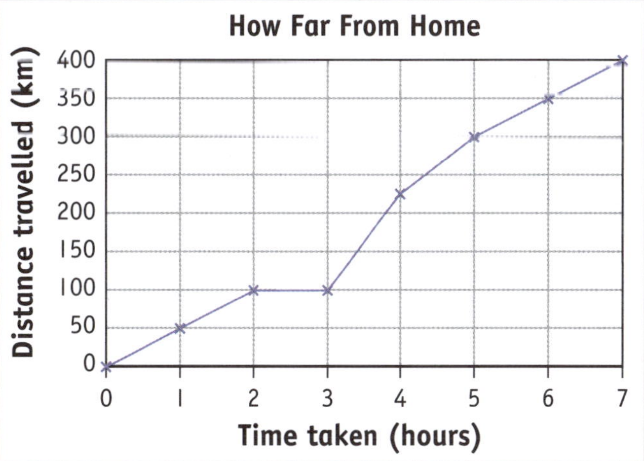

Answer these questions to tell the story of the journey.

1. How far did the family travel that day?

   _______ km

2. For how much time were they on the road?

   ___ hours

3. In which hour did they travel the furthest? Between hours ___ and ___
4. Why were they stationary for 1 hour? ___________________________
5. Why did they travel more slowly in the first two hours than during the rest of the time that they were driving?

   ____

SCORE /5 0-2 3-4 5

TARGETING MATHS HOMEWORK: YEAR 5 © PASCAL PRESS ISBN: 9781925726589

TERM 4

AC9M5M04

Angles of a Quadrilateral

There is something special about the angles inside a quadrilateral. Measure each of the angles in these quadrilaterals and then explain your findings.

①

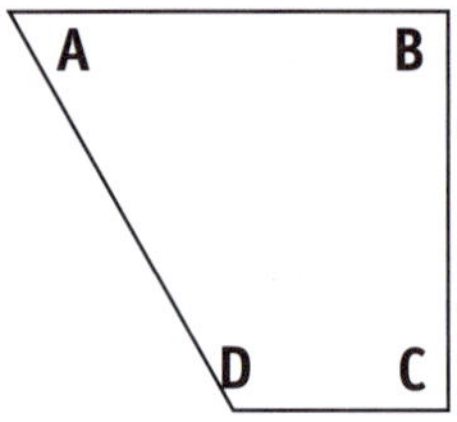

____° + ____° + ____° + ____° = ____°

③

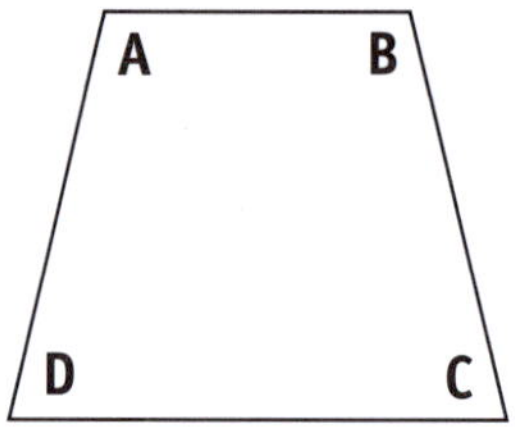

____° + ____° + ____° + ____° = ____°

②

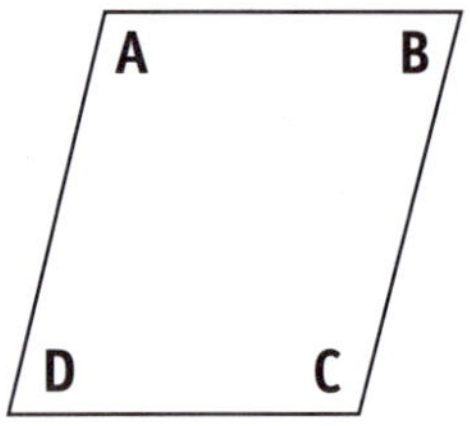

____° + ____° + ____° + ____° = ____°

④

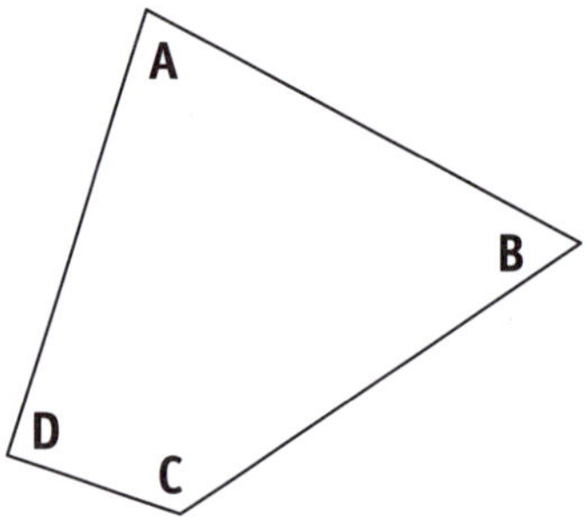

____° + ____° + ____° + ____° = ____°

⑤ What did you find out about the internal angles of a quadrilateral?

____

SCORE /5 0-2 3-4 5

NAPLAN*-style Questions

① Which of these numbers is a multiple of 6 and has 4 as one of its factors?

- ◯ 9
- ◯ 16
- ◯ 18
- ◯ 36

②

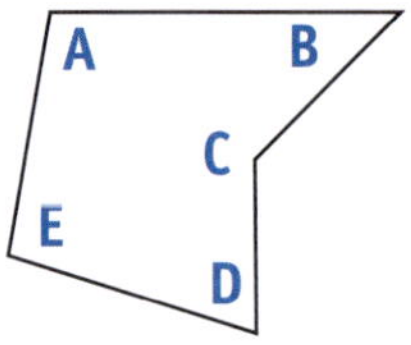

This shape has five angles marked.

Which angle is 45°?

Angle ▢

③ The line graph shows the time taken by a competitor in the Fun Run.

Which statement about the graph could be true?

- ◯ The competitor ran at the same speed all the time.
- ◯ The Fun Run was uphill for the first half of the race.
- ◯ The runner ran fastest during the second hour of the race.

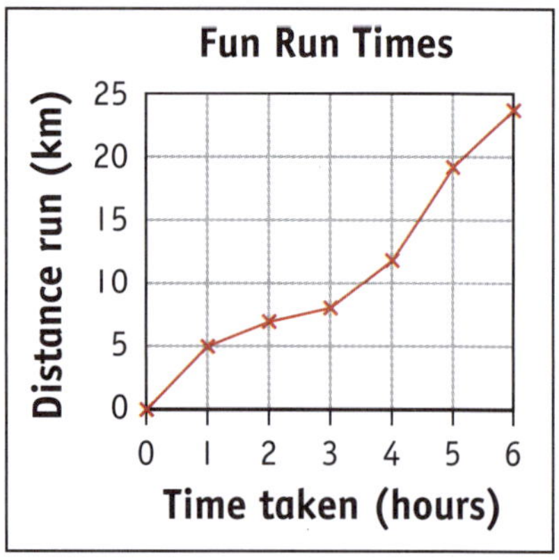

SCORE /3 0-1 2 3

TERM 4

TARGETING MATHS HOMEWORK: YEAR 5 © PASCAL PRESS ISBN: 9781925726589

Problem of the Week

AC9M5N08

Factors and Multiples on a Hexagon

Sam and Ravi have been making factor–multiple hexagons.

Their rule is to start with any number you like, but it has to be a multiple of the next number on the hexagon, which then has to be a factor of the next number on the hexagon.

Here is their first attempt.

"We thought we would use the numbers that Asha used: 2, 3, 4, 6, 8 and 12."

We started with **4 is a multiple of 2**: Then we added more numbers:

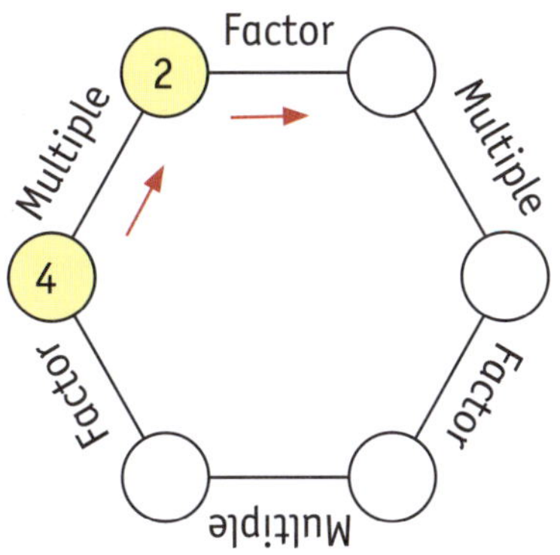

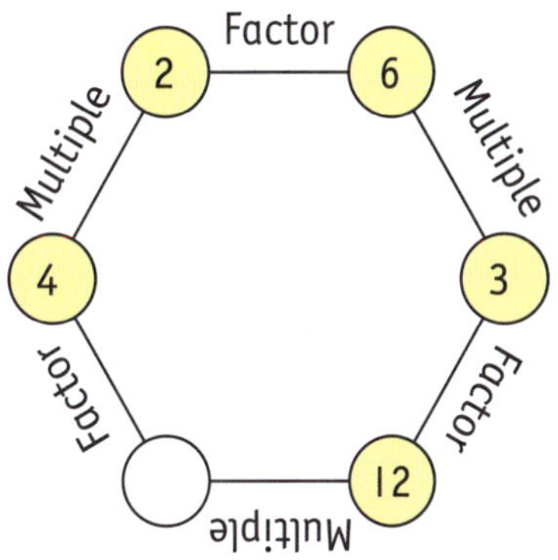

"Then we were stuck! The number 8 is not a multiple of 12 or a factor of 4."

The rules that Sam and Ravi made were that you have to start and finish with the same number, and the numbers in between must all be different.

The Challenge

You might like to start with finding a correct way to use Asha's numbers before you go on to make more factor–multiple hexagons.

Also, some of the sequences on page 111 would be a good place to start.

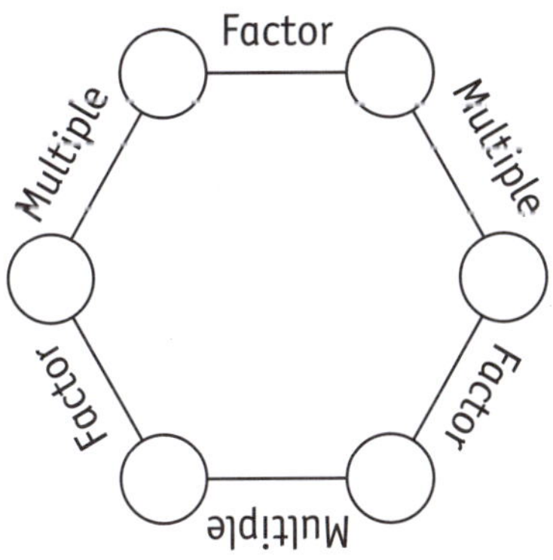

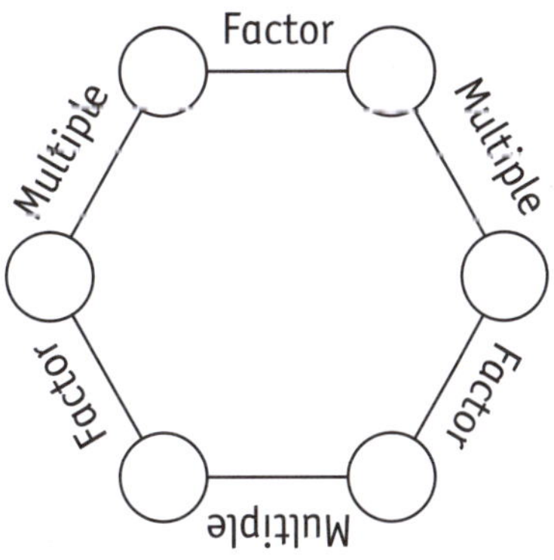

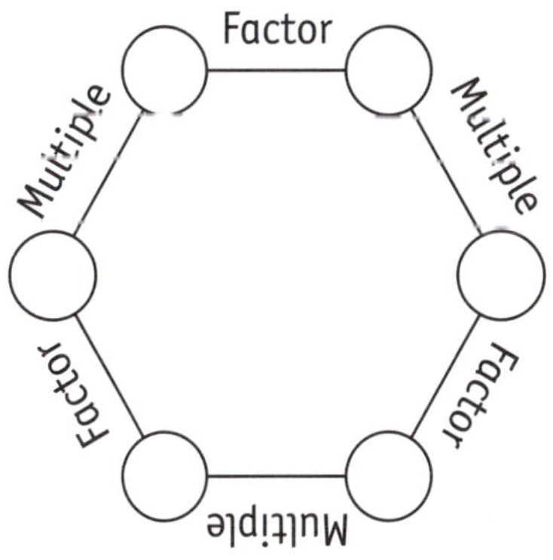

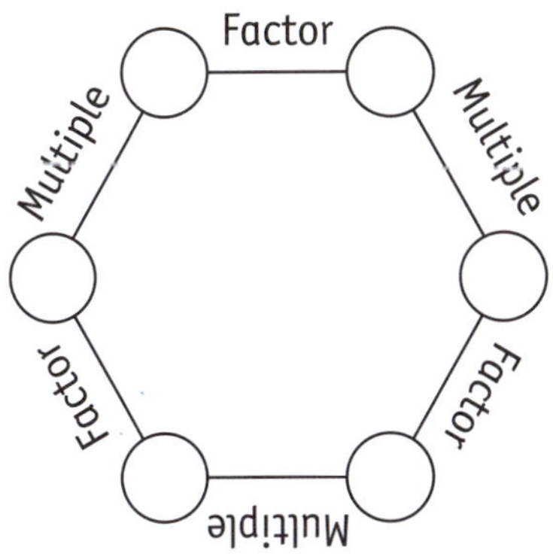

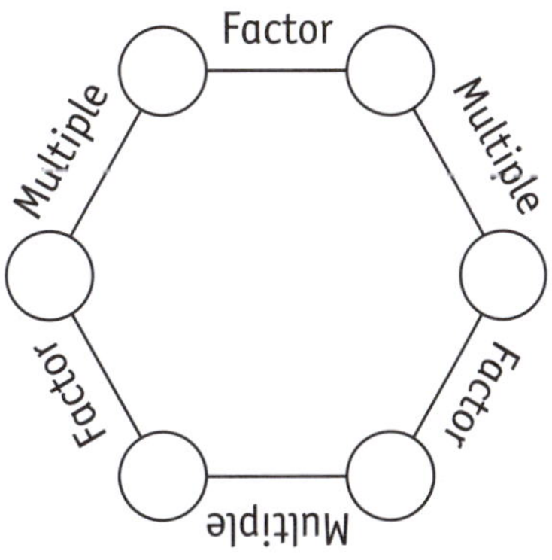

Did you find some sequences of numbers that are different from the ones that were used on page 111?
Well done if you did!

AC9M5N03

Comparing Fractions

To put the fractions $\frac{3}{8}$, $\frac{3}{4}$ and $1\frac{1}{2}$ in order on a number line,
convert them into equivalent fractions with the same denominator.
For this example, the common denominator is 8, so convert all the fractions into eighths.

$\frac{3}{8}$	$\frac{3}{4} = \frac{6}{8}$	$1\frac{1}{2} = \frac{3}{2} = \frac{12}{8}$
No change needed.	Multiply the numerator and the denominator by 2.	Multiply the numerator and the denominator by 4.

Change these fractions into equivalent fractions with the same denominator. Then place them in order on the number line.

1. $1\frac{1}{8}$ = $\frac{3}{4}$ = $1\frac{1}{2}$ =

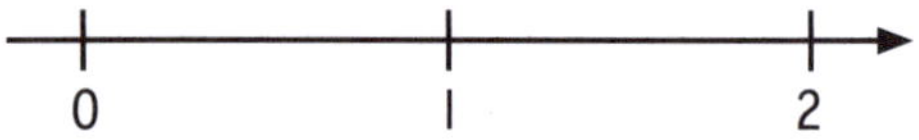

2. $\frac{3}{3}$ = $\frac{5}{6}$ = $\frac{4}{3}$ =

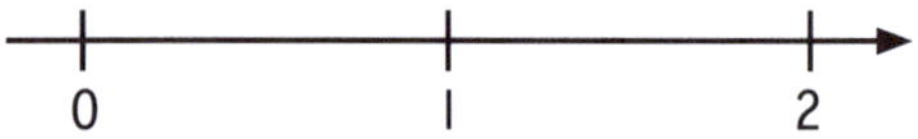

3. $\frac{3}{5}$ = $1\frac{4}{5}$ = $\frac{13}{10}$ =

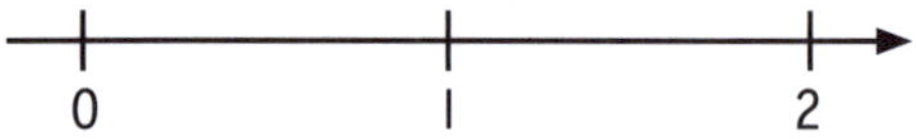

4. $\frac{3}{8}$ = $\frac{4}{16}$ = $1\frac{1}{8}$ =

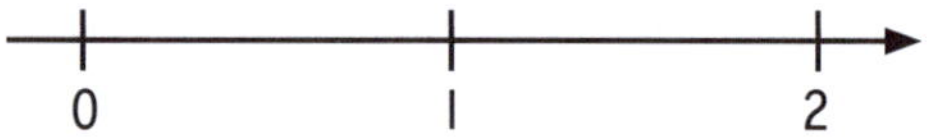

5. $\frac{5}{6}$ = $\frac{4}{12}$ = $1\frac{1}{6}$ =

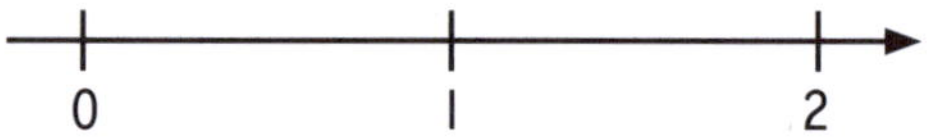

6. $\frac{5}{4}$ = $\frac{5}{8}$ = $1\frac{3}{8}$ =

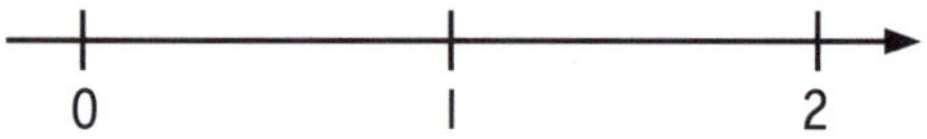

SCORE /6 0-2 3-4 5-6

Quick Check

**Write the fraction marked with ? on these number lines.
Read the end number on each number line very carefully.**

1.

2.
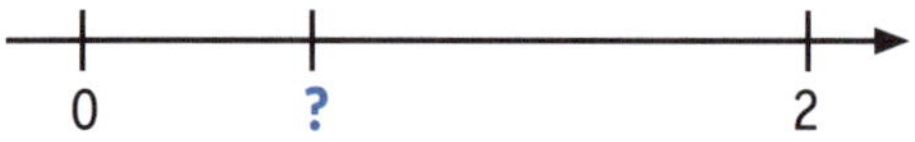

3.
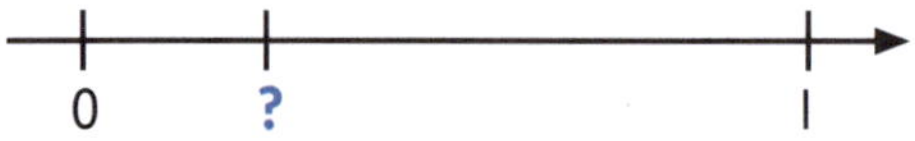

4.
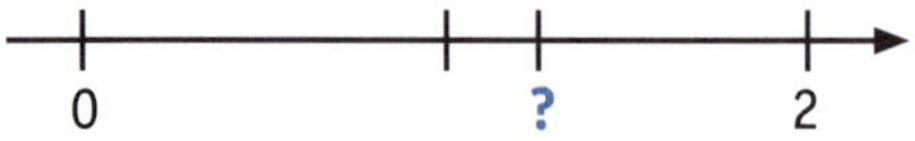

5.
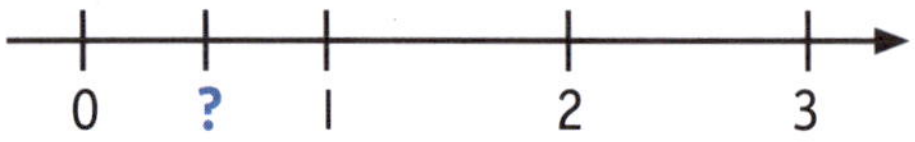

6.

7.
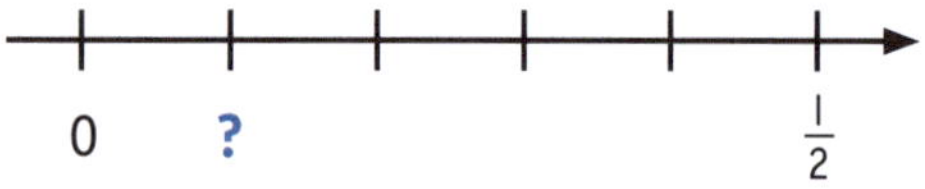

8.
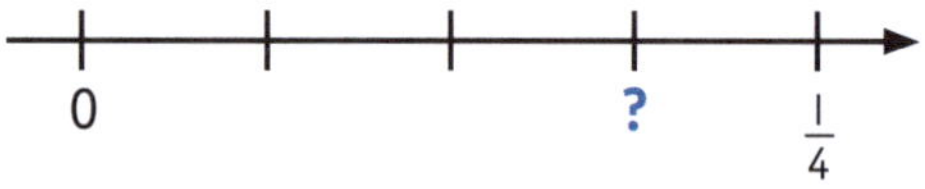

9.
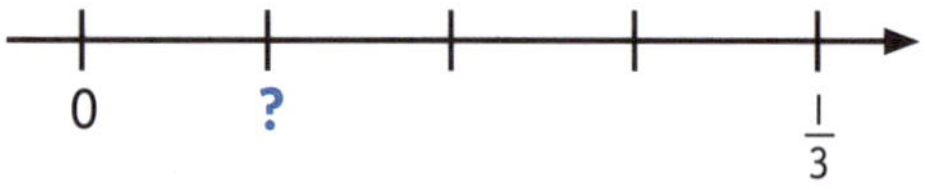

10.
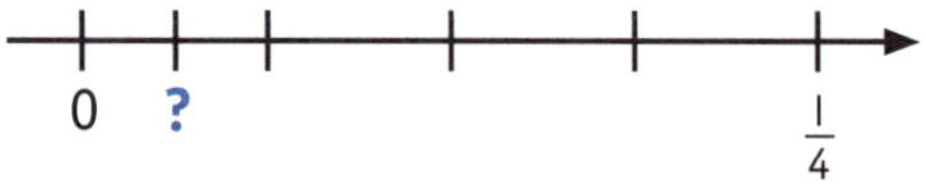

SCORE /10 0-4 5-8 9-10

TERM 4

TARGETING MATHS HOMEWORK: YEAR 5 © PASCAL PRESS ISBN: 9781925726589

Number & Algebra

AC9M5N04, AC9M5N07

Decimals to Fractions

As you know:

- 50% is equivalent to $\frac{1}{2}$. So we can find 50% of amounts by dividing by two.
- 25% is equivalent to $\frac{1}{4}$. So we can find 25% of amounts by dividing by 4.

Use halves and quarters as your strategy to work out these percentages.

1. 50% of $70 = ________
2. 25% of $100 = ________
3. 25% of $160 = ________
4. 50% of $1 = ________
5. 75% of $100 = ________
6. 25% of $25 = ________
7. 50% of $50 = ________
8. 75% of $180 = ________
9. 25% of $96 = ________
10. 50% of $5000 = ________

SCORE /10 0-4 5-8 9-10

Statistics & Probability

AC9M5P02

Design a Spinner

Jade, Miro and Ella have been designing spinners.
The spinners have different shapes and will have red, yellow and blue segments.

Complete the spinners for Jade, Miro and Ella.

1. Jade says:
"I want my spinner to have a $\frac{1}{3}$ chance of yellow winning and I want blue to have the least chance."

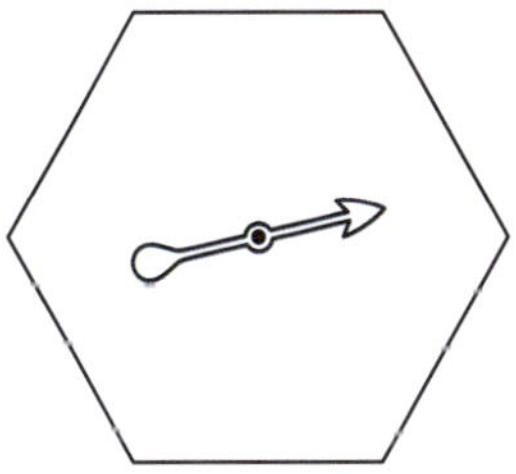

2. Miro says:
"I want red and yellow to have equal chances of winning and a better chance of winning than blue."

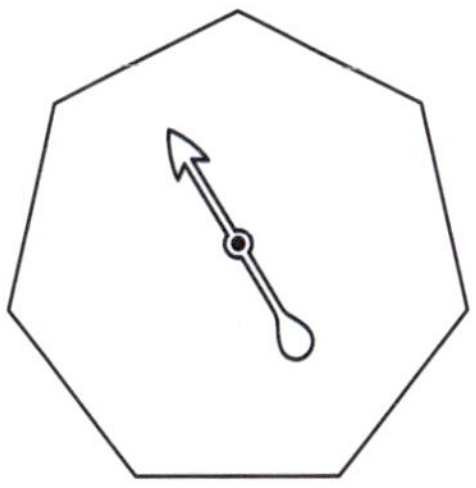

3. Ella says:
"I want my spinner to have a 50% chance of red winning and for yellow to have a better chance of winning than blue."

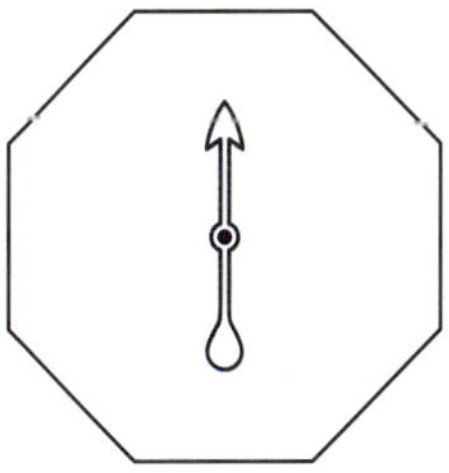

SCORE /3 1 2 3

AC9M5SP03

Transformations

1. Draw this shape after it has been translated 4 grid squares to the right.

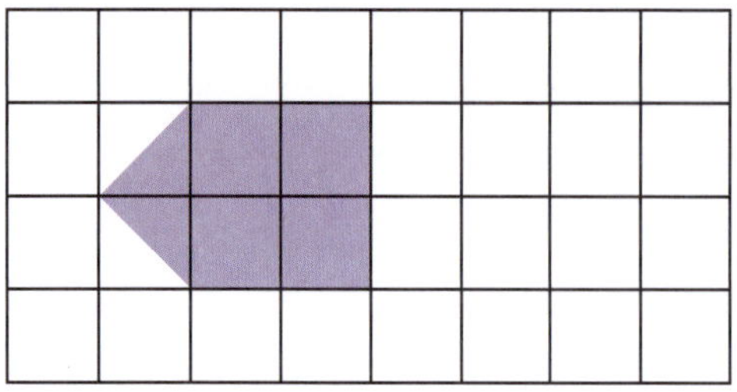

2. Draw this shape after it has been rotated 90° clockwise about the point marked with a dot.

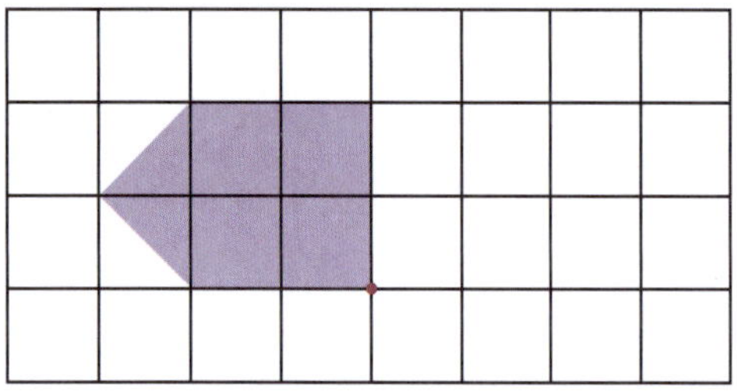

3. Draw this shape after it has been reflected in the dashed line.

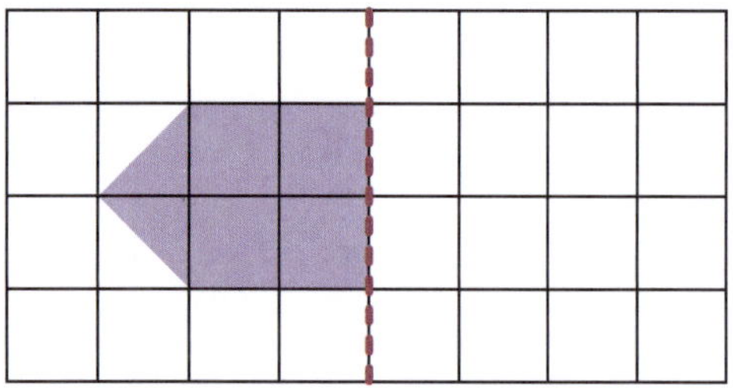

4. Mark the lines of symmetry in the combined shapes that you have made.

5. Which shape has two lines of symmetry?

   Question ___

6. Which shape has no lines of symmetry?

   Question ___

7. Draw this shape on the blank grid below but enlarge it to twice its size.

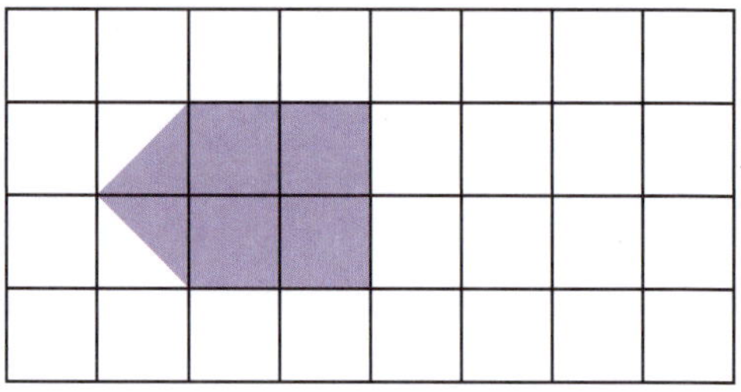

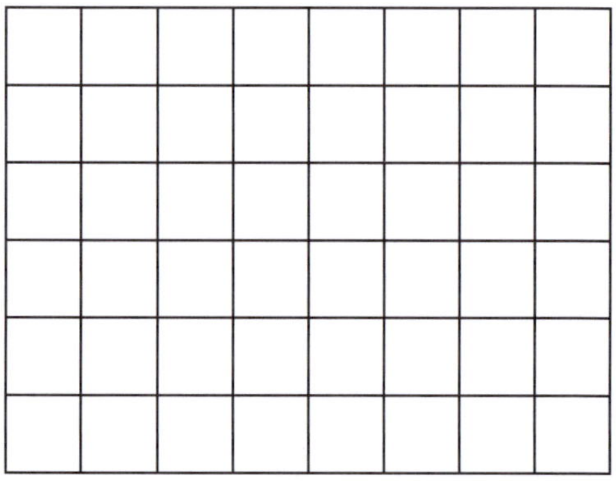

SCORE /7 (0-3) (4-5) (6-7)

NAPLAN*-style Questions

1. Which of these fractions is the smallest?

   ◯ $\frac{1}{4}$ ◯ $\frac{3}{8}$ ◯ $\frac{3}{16}$ ◯ $\frac{2}{12}$

2. Which spinner has a 50% chance of spinning white?

   ◯ 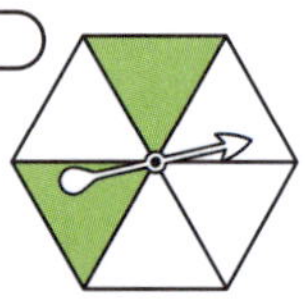 ◯ 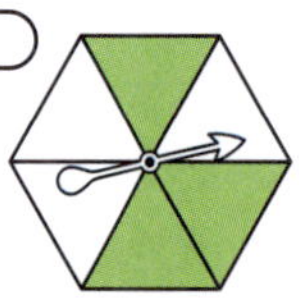 ◯ 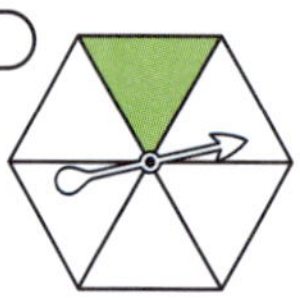 ◯ 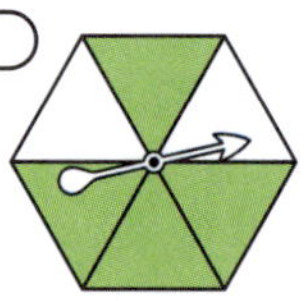

3. Which transformation has been made to this shape?

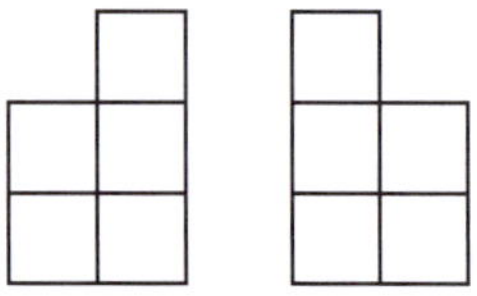

   ◯ Translation
   ◯ Rotation
   ◯ Reflection

SCORE /3 (0-1) (2) (3)

TARGETING MATHS HOMEWORK: YEAR 5 © PASCAL PRESS ISBN: 9781925726589

Race to 1

Will, Asha and Ravi have invented a fraction game called Race to 1.

We made up this board with fraction number lines to help us play the game.

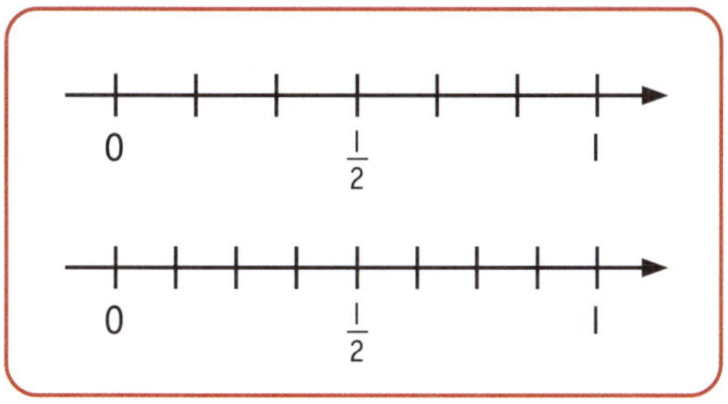

These are the cards that we made for the game.

$\frac{1}{2}$ $\frac{1}{3}$ $\frac{2}{3}$ $\frac{1}{4}$ $\frac{2}{4}$ $\frac{3}{4}$

$\frac{1}{6}$ $\frac{2}{6}$ $\frac{3}{6}$ $\frac{4}{6}$ $\frac{5}{6}$

$\frac{1}{8}$ $\frac{2}{8}$ $\frac{3}{8}$ $\frac{4}{8}$ $\frac{5}{8}$ $\frac{6}{8}$ $\frac{7}{8}$

We keep them in a bag, and we give the bag a good shake before the start of each game.

To play, we take turns to pick one card from the bag.

Each round we add one more card to our collection.

We can use any of our cards to make jumps along either of the number lines.

The first person who has cards that can make jumps from 0 to 1 is the winner!

Here's a simple example of three cards that you can use to make jumps from 0 to 1.

$\frac{3}{6}$ $\frac{1}{3}$ $\frac{1}{6}$ →

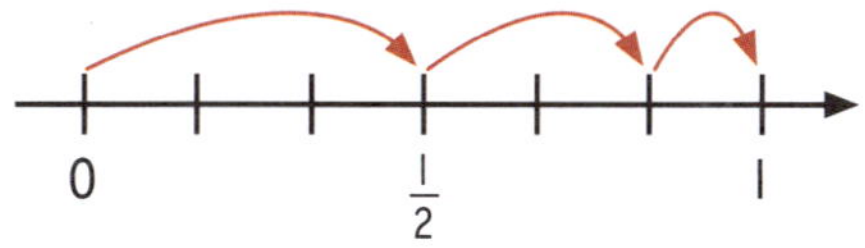

The Challenge

Will, Asha and Ravi have played three rounds and these are the cards that they hold:

Will has: $\frac{2}{3}$ $\frac{3}{4}$ $\frac{5}{6}$

Asha has: $\frac{1}{3}$ $\frac{2}{8}$ $\frac{7}{8}$

Ravi has: $\frac{1}{8}$ $\frac{2}{4}$ $\frac{1}{6}$

1. What cards does each player need in Round 4 to win the **Race to 1**?

   Will needs ________________. Asha needs ________________. Ravi needs ________________.

2. Who has the best chance of becoming the winner after Round 4? __________

TERM 4

Adding Fractions

Adding strings of fractions can look pretty hard at first glance:

$$\frac{1}{2} + \frac{1}{4} + \frac{1}{8} + \frac{3}{4} + \frac{2}{6} + \frac{1}{6}$$

But if you look for **compatible** parts, it's not so bad.

$\frac{2}{6} + \frac{1}{6} = \frac{3}{6} = \frac{1}{2}$ so that compatible pair is also compatible with $\frac{1}{2}$.

$\frac{1}{4} + \frac{3}{4}$ are a compatible pair that make 1.

This diagram uses colour to show which fractions are compatible and can be grouped together:

$$\boxed{\frac{1}{2}} + \boxed{\frac{1}{4}} + \frac{1}{8} + \boxed{\frac{3}{4}} + \boxed{\frac{2}{6} + \frac{1}{6}} = 2\frac{1}{8}$$

Use colour to show compatible fractions in these fraction additions and then write the answers. You will need to use equivalent fractions too, as in $\frac{3}{5}$ is equivalent to $\frac{6}{10}$.

1. $\frac{2}{5} + \frac{2}{3} + \frac{1}{4} + \frac{6}{10} + \frac{2}{6} =$
2. $\frac{1}{3} + \frac{1}{8} + \frac{1}{2} + \frac{1}{4} + \frac{4}{8} + \frac{2}{3} =$
3. $\frac{3}{5} + \frac{5}{6} + \frac{1}{2} + \frac{1}{6} + \frac{1}{4} + \frac{2}{5} =$
4. $\frac{1}{4} + \frac{3}{5} + \frac{2}{5} + \frac{3}{4} + \frac{1}{2} + \frac{1}{6} =$
5. $\frac{3}{10} + \frac{1}{2} + \frac{1}{5} + \frac{5}{8} + \frac{1}{6} + \frac{3}{8} =$
6. $\frac{1}{3} + \frac{1}{4} + \frac{1}{6} + \frac{2}{3} + \frac{2}{4} + \frac{2}{6} =$
7. $\frac{1}{10} + \frac{1}{5} + \frac{2}{10} + \frac{2}{5} + \frac{3}{10} + \frac{3}{5} + \frac{6}{10} =$

SCORE /7 0-3 4-5 6-7

TERM 4

Quick Check

As quickly as you can, find the compatible parts that add to 1 in these fraction additions and then write the answer.

1. $\frac{3}{4} + \frac{1}{2} + \frac{1}{4} =$
2. $\frac{3}{5} + \frac{3}{10} + \frac{2}{5} =$
3. $\frac{7}{8} + \frac{1}{2} + \frac{2}{16} =$
4. $\frac{3}{8} + \frac{1}{4} + \frac{5}{8} =$
5. $\frac{4}{10} + \frac{3}{5} + \frac{1}{6} =$
6. $\frac{1}{3} + \frac{1}{4} + \frac{3}{8} + \frac{3}{8} =$
7. $\frac{2}{5} + \frac{3}{6} + \frac{1}{3} + \frac{1}{6} =$
8. $\frac{3}{10} + \frac{1}{2} + \frac{1}{4} + \frac{2}{10} =$
9. $\frac{1}{4} + \frac{2}{3} + \frac{6}{8} + \frac{1}{6} =$
10. $\frac{5}{8} + \frac{3}{4} + \frac{3}{8} + \frac{1}{4} =$

SCORE /10 0-4 5-8 9-10

TARGETING MATHS HOMEWORK: YEAR 5 © PASCAL PRESS ISBN: 9781925726589

Number & Algebra

AC9M5N06

Multiplication Is Commutative

2 × 3 × 5 × 6

You can multiply these numbers in any order, and you always get the same result.

For example:

2 × 6 = 12, 12 × 5 = 60, 60 × 3 = 180

6 × 5 = 30, 30 × 3 = 90, 90 × 2 = 180

3 × 6 = 18, 18 × 5 = 90, 90 × 2 = 180

Multiplication is **commutative**, which means that you can change the order in which you work out the multiplications and you will always get the same result.

Use the commutative property to multiply these numbers by choosing an order for multiplying that makes it easy for you to find the answer.

1. 2 × 3 × 4 × 7 ____ × ____ = ____, ____ × ____ = ____, ____ × ____ = ____
2. 4 × 6 × 5 × 3 ____ × ____ = ____, ____ × ____ = ____, ____ × ____ = ____
3. 2 × 7 × 4 × 5 ____ × ____ = ____, ____ × ____ = ____, ____ × ____ = ____
4. 6 × 5 × 8 × 3 ____ × ____ = ____, ____ × ____ = ____, ____ × ____ = ____
5. 3 × 5 × 2 × 9 ____ × ____ = ____, ____ × ____ = ____, ____ × ____ = ____

SCORE /5 0-2 3-4 5

Statistics & Probability

AC9M5ST01

Borrowed Books

The school librarian was asked to present a column graph to show which types of books were borrowed from the library last term.

Results:

120 Fiction, 80 Animal books, 30 How To books, 80 Joke books, 40 Picture books

1. Use the grid to make a column graph for the librarian. Think carefully about the scale to use.
2. What is the mode of the data? ____
3. How many books were borrowed last term? ____
4. Which two categories together had the same number of books borrowed as Fiction?

   ____________ and ____________
5. How many more Joke and Animal books were borrowed than Fiction books?

   ____

SCORE /5 0-2 3-4 5

TERM 4

AC9M5M01

Designing a City Skyline

This is Jade's design for a city skyline:

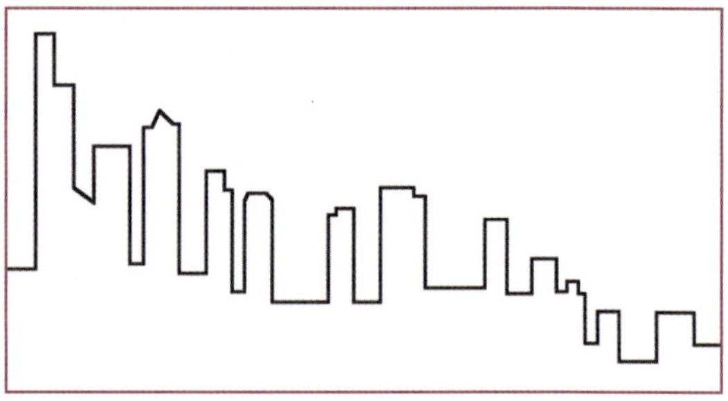

Follow these instructions in the box below to make your own city skyline.

1. Start at the red dot and draw a horizontal line 140 mm long. Mark the end of the line with a cross.
2. Draw a perpendicular line 4 cm in length upwards from the red dot.
   Note: Perpendicular means at a 90 degree angle to the horizontal line.
3. Draw a perpendicular line 6 cm in length upwards from the cross at the end of the line.
4. Create a skyline with at least two buildings that joins the two vertical lines you drew.
5. List each of the vertical measurements that you made: ____________________
6. List each of the horizontal measurements that you made: ____________________

SCORE /6 0-2 3-4 5-6

NAPLAN*-style Questions

1. What is the missing fraction in this equation?

   $\frac{1}{4} + \square = \frac{3}{6} + \frac{1}{2}$

2. Plant 1 is 140 centimetres tall.
   About how much taller is Plant 1 than Plant 2?

   [ ] cm

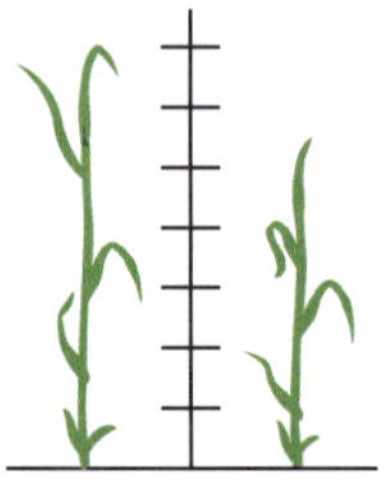

3. The canteen ordered 5 packets of 6 round bread rolls and 4 packets of 4 long bread rolls.
   How many bread rolls did the canteen order altogether?

   [ ] bread rolls

SCORE /3 0-1 2 3

TARGETING MATHS HOMEWORK: YEAR 5 © PASCAL PRESS ISBN: 9781925726589

TERM 4

Problem of the Week

UNIT 27

AC9M5M01

Estimate and Check

Mrs Green gave each student a piece of string, a ruler and a set of four letters that were each drawn inside rectangles the same size.

"Your challenge for today," she said, "is to **estimate** the length of the lines used to make the letters and find a way to measure the lines as a **check** of the estimate.

"After that, use the blank squares to draw in four different letters from your name. Your letters must touch all four sides, just like mine do."

Challenge 1

Estimate how long the lines of each letter are and then measure to check.

1. Estimate: ____
   Check: ____
2. Estimate: ____
   Check: ____
3. Estimate: ____
   Check: ____
4. Estimate: ____
   Check: ____

Challenge 2

Choose four letters from your own name and draw the letters in the squares below.
Make sure that the lines you draw touch the sides of the square for each letter.
Then estimate and check the line lengths.

5. Estimate: ____
   Check: ____
6. Estimate: ____
   Check: ____
7. Estimate: ____
   Check: ____
8. Estimate: ____
   Check: ____

TERM 4

Division

Now that you know about **factors** and **rules for divisibility**, you can use both to simplify division questions.

For 189 ÷ 36, we know that both numbers have 9 as a factor because 1 + 8 + 9 = 18 which is a multiple of 9, and 3 + 6 = 9 which is also a multiple of 9.

So, to find 189 ÷ 36, we can follow the steps shown in the diagram:

189 ÷ 36
÷ 9 ↓ ↓
21 ÷ 4
5 r 1

Complete these division questions using what you know about factors and rules of divisibility.

1. 456 ÷ 88
   ÷ 8 ↓ ↓
   57 ÷ ___
   ___ r ___

2. 372 ÷ 12
   ÷ 6 ↓ ↓
   ___ ÷ ___
   ___ r ___

3. 584 ÷ 72
   ÷ ___ ↓ ↓
   ___ ÷ ___
   ___ r ___

4. 488 ÷ 32
   ÷ ___ ↓ ↓
   ___ ÷ ___
   ___ r ___

5. 574 ÷ 35
   ÷ ___ ↓ ↓
   ___ ÷ ___
   ___ r ___

6. 364 ÷ 28
   ÷ ___ ↓ ↓
   ___ ÷ ___
   ___ r ___

SCORE /6 (0-2) (3-4) (5-6)

Quick Check

Remember: a number is divisible by 4 if its last two digits are a multiple of 4.

Complete these Yes/No questions.

Is this number divisible by 9?

1. 173 Yes/No
2. 261 Yes/No
3. 434 Yes/No
4. 521 Yes/No
5. 627 Yes/No

Is this number divisible by 4?

6. 356 Yes/No
7. 224 Yes/No
8. 372 Yes/No
9. 436 Yes/No
10. 214 Yes/No

SCORE /10 (0-4) (5-8) (9-10)

TARGETING MATHS HOMEWORK: YEAR 5 © PASCAL PRESS ISBN: 9781925726589

Multiplication

You can use **factors** to simplify multiplication questions.
Now that you know multiplication is **commutative**, you know that you can multiply numbers in any order and the answer remains the same.
To work out 15 × 16, you can break down the 15 and 16 into their factors:
Now you can choose the easiest order in which to multiply.
This might be 8 × 5 × 3 × 2, so 8 × 5 = 40, 40 × 3 = 120 and 120 × 2 = 240.

15 × 16
3 × 5 × 2 × 8

Use factors for these multiplications.

1. 24 × 18
= ___ × ___ × ___ × ___
___ × ___ × ___ × ___ = _____

2. 15 × 14
= ___ × ___ × ___ × ___
___ × ___ × ___ × ___ = _____

3. 18 × 12
= ___ × ___ × ___ × ___
___ × ___ × ___ × ___ = _____

4. 24 × 12
= ___ × ___ × ___ × ___
___ × ___ × ___ × ___ = _____

5. 24 × 16
= ___ × ___ × ___ × ___
___ × ___ × ___ × ___ = _____

6. 14 × 14
= ___ × ___ × ___ × ___
___ × ___ × ___ × ___ = _____

SCORE /6 0-2 3-4 5-6

Statistics & Probability

AC9M5ST01

Reading a Chart

This graph shows the number of adult and child tickets sold for a puppet show.

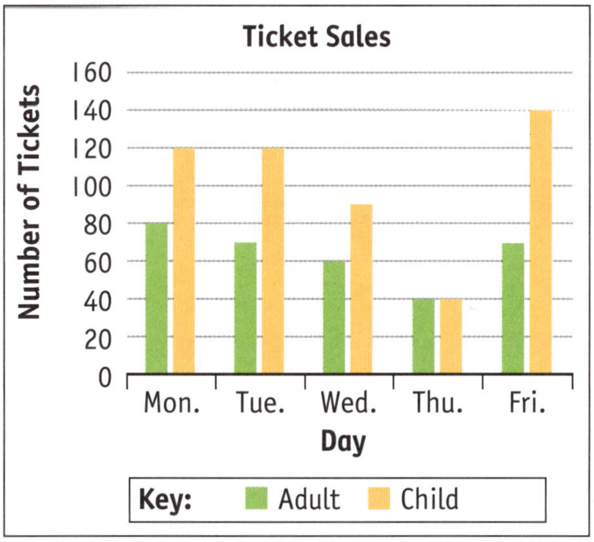

1. On which day were the most adult tickets sold?
____

2. On which day were 60 adult tickets sold?
____

3. On which day were the least adult tickets sold?
____

4. What is the mode for the number of child tickets sold?
____ tickets

5. How many more child tickets than adult tickets were sold on Friday? ____

SCORE /5 0-2 3-4 5

TERM 4

AC9M5M01, AC9M5M02

Area and Perimeter

Ella and Will have been drawing shapes on grid paper and investigating their areas and perimeters.

"We discovered that comparing different areas and perimeters is not quite as easy as we thought it would be."

Here are pictures of four of the shapes that they drew on 1 cm grid paper.

Shape A | Shape B | Shape C | Shape D

Before checking

① Estimate which shape has the shortest perimeter.

Shape ____

② Estimate which shape has the largest area.

Shape ____

Now check your estimates.

③ Use a ruler to check your perimeter estimate. Which shape has the shortest perimeter?

Shape ____

④ Count the grid squares to check your area estimate. Which shape has the largest area?

Shape ____

SCORE /4 0-1 2-3 4

NAPLAN*-style Questions

① The school has collected $196 to be shared equally to buy 8 raffle prizes. How much can be spent on each prize?

$ __________

② Which of these numbers cannot be divided evenly by 9?

◯ 183 ◯ 279 ◯ 324 ◯ 263

③ Which of the shapes below has the same area as this green one?

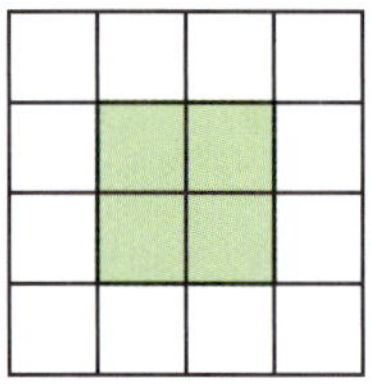

◯ 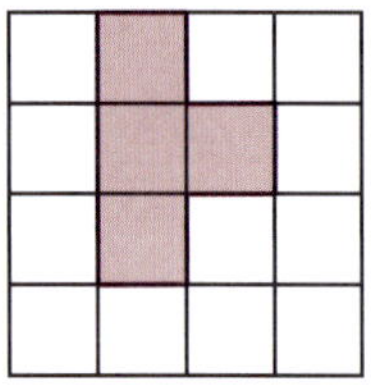

◯ 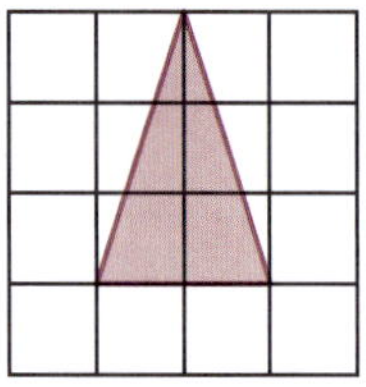

◯ 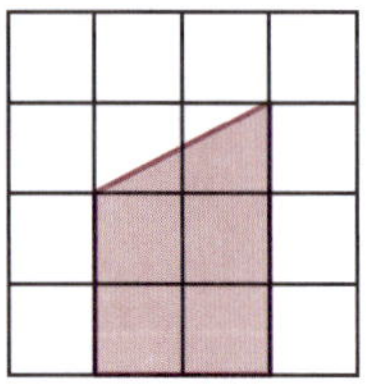

◯ 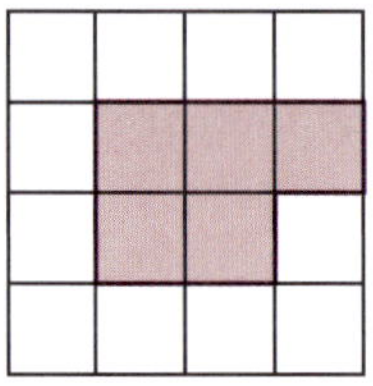

SCORE /3 0-1 2 3

TARGETING MATHS HOMEWORK: YEAR 5 © PASCAL PRESS ISBN: 9781925726589

Making 2-digit Numbers

Asha and Ravi have set themselves a puzzle and they would like your help to find answers.

We have written the numbers 1 to 9 on cards.

1	2	3	4	5	6	7	8	9

We shuffle the cards and then deal out four cards to play with.

Asha and Ravi use the four cards to make two 2-digit numbers and multiply them to make an answer.

☐☐ × ☐☐ = ______

Asha has to arrange the cards to make the answer as large as possible.

Ravi has to arrange the cards to make the answer as small as possible.

Challenge 1

Asha and Ravi have dealt the numbers **2**, **3**, **5** and **7**.

What should their multiplications be?

Asha's arrangement should be:

☐☐ × ☐☐ = ______

Ravi's arrangement should be:

☐☐ × ☐☐ = ______

Challenge 2

Asha and Ravi have dealt the numbers **3**, **5**, **7** and **8**.

What should their multiplications be?

Asha's arrangement should be:

☐☐ × ☐☐ = ______

Ravi's arrangement should be:

☐☐ × ☐☐ = ______

Describe how both players should arrange their numbers, no matter what cards have been dealt.

____

Challenge 3

What if the multiplication had to be like this?

☐☐☐ × ☐ = ______

What would be the results for the numbers dealt in Challenge 1 and Challenge 2?

____

____

TERM 4

Decimal Multiplication

The baker was making a spreadsheet to show what each of his items would cost when a customer bought one or more of them.

Number	Bread Rolls	Donuts	Muffins	Sausage Rolls
1	$1·15	$2·35	$1·85	$3·35
2				$6·70
3				$10·05
4				$13·40
5	$5·75	$11·75	$9·25	$16·75

Suddenly the computer malfunctioned and he was unable to complete the spreadsheet.

1. Use a calculator to fill in the missing values in the baker's spreadsheet.
2. He has an order for 6 sausage rolls and 4 bread rolls. How much will that cost?

   $____________
3. He has an order for 32 muffins. How much will that cost?

   $____________
4. He has an order for 15 donuts and 15 muffins. How much will that cost?

   $____________
5. He has an order for 12 of everything. How much will that cost?

   $____________
6. He has an order for 12 bread rolls and 18 sausage rolls. How much will that cost?

   $____________

TERM 4

SCORE /6 0-2 3-4 5-6

Quick Check

Write the cost of each of these bakery orders.

1. 8 bread rolls = $________
2. 7 muffins = $________
3. 10 donuts = $________
4. 11 sausage rolls = $________
5. 9 donuts = $________
6. 1 of everything = $________
7. 10 of everything = $________
8. 7 donuts and 7 muffins = $________
9. 20 sausage rolls = $________
10. 40 muffins = $________

SCORE /10 0-4 5-8 9-10

TARGETING MATHS HOMEWORK: YEAR 5 © PASCAL PRESS ISBN: 9781925726589

Number & Algebra

Lowest Common Multiples

When you have two numbers to work with, it is interesting to find a third number that they are both factors of.

If you start with 6 and 15, the third number could be 90, but that's not the smallest possibility.

Look at the factors of the two starting numbers:

6 → 2 × **3**   15 → **3** × 5

You can see that 3 is a repeated factor and we don't need to use it twice, so:

6, 15 → 2 × **3** × 5 = 30

So, **30** is the smallest number that has both **6 and 15** as factors. It is called the **lowest common multiple** (LCM).

List the factors of the number pairs, show which is a repeated factor and find their lowest common multiple.

1. What is the lowest common multiple of 6 and 8?

   6   8

   ___ × ___ × ___ = _____

2. What is the lowest common multiple of 15 and 21?

   15   21

   ___ × ___ × ___ = _____

3. What is the lowest common multiple of 12 and 16?

   12   16

   ___ × ___ × ___ = _____

4. What is the lowest common multiple of 18 and 24?

   18   24

   ___ × ___ × ___ = _____

5. What is the lowest common multiple of 6 and 14?

   6   14

   ___ × ___ × ___ = _____

SCORE /5 0-2 3-4 5

Statistics & Probability

AC9M5ST02

Morning Temperatures

Leo made this chart of morning temperatures last Monday and Tuesday.

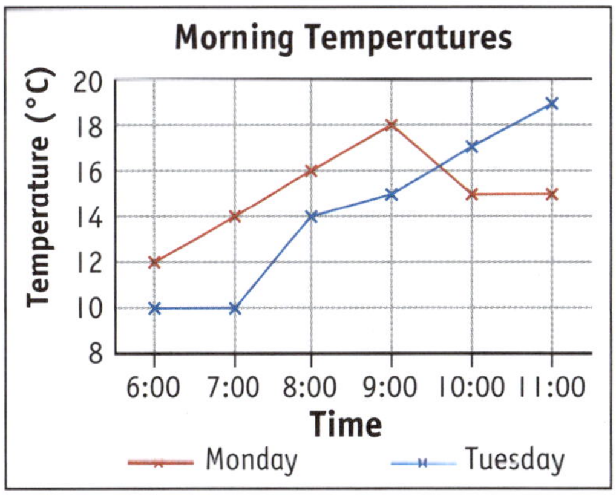

1. By how much did the temperature increase on Monday morning between 6 am and 9 am? _____ °C
2. By how much did the temperature increase on Tuesday morning between 8 am and 11 am? _____ °C
3. When was the greatest temperature increase in 1 hour?

   On ________________ between ________ and ________

4. On Monday, by how much did the temperature drop after 9:00 am?

   _____ °C

5. Why might the temperature have dropped by that much on Monday?

   ____

SCORE /5 0-2 3-4 5

TERM 4

AC9M5M03

Time Zones

Time zones around the world are not the same and they have different names.

- London runs on Greenwich Mean Time (GMT).
- Perth runs on Western Standard Time (WST).
- Sydney runs on Eastern Standard Time (EST).

This table shows the time in each of those time zones.
You can see that Sydney is ahead of Perth and London.

Times in London, Perth and Sydney		
London (GMT)	**Perth (WST)**	**Sydney (EST)**
7:30 am	3:30 pm	6:30 pm (same day)
Midday	8:00 pm	11:00 pm (same day)
4:00 pm	Midnight	3:00 am (next day)
Midnight	8:00 am	11:00 am (same day)

Write the correct time and whether it is on the same day, the next day or the previous day.
Note: Midnight is the start of a day and any time before that is on the previous day.

1. At 12:30 am in London, it is ________ in Sydney on the __________ day.
2. At midnight in Sydney, it is ________ in Perth on the __________ day.
3. At 5:00 am in Sydney, it is ________ in London on the __________ day.
4. At 2:00 pm in London, it is ________ in Perth on the __________ day.
5. At 10:00 pm in Perth, it is ________ in London on the __________ day.
6. At 3:15 pm in London, it is ________ in Sydney on the __________ day.
7. At noon in Sydney, it is ________ in London on the __________ day.
8. At 3:45 am in Perth, it is ________ in London on the __________ day.

SCORE /8 (0-3) (4-6) (7-8)

NAPLAN*-style Questions

1. Select the numbers that have 6 as a factor.

   ◯ 153 ◯ 132 ◯ 406 ◯ 1344 ◯ 252 ◯ 372 ◯ 560

2. Which number sentence matches this question?
   Jade bought six notebooks at $3·90 each. How much did she spend?

   ◯ $6 + $3·90 ◯ 6 × $4 − 6 × 10c ◯ 6 × $3 + 90c

3. What is the mode of these numbers?

   3, 6, 6, 6, 3, 4, 3, 6

   Mode: ☐

SCORE /3 (0-1) (2) (3)

TARGETING MATHS HOMEWORK: YEAR 5 © PASCAL PRESS ISBN: 9781925726589

Problem of the Week

AC9M5N08, AC9M5N09

Morning Tea Breaks

The Sports Club put on morning breaks for their helpers.
There can be between 3 and 6 helpers on any one day.
They order food from the local coffee shop:

Sausage Roll	$2·50
Cheese Skewer	$3·10
Fruit Skewer	$2·75
Muffin	$3·30
Ham Roll-up	$3·90

Use the menu to work out the answers. The items don't have to be all different, but the total price must match what the Sports Club spent.

1. On Day 1, the Sports Club spent $8 and bought 3 items for 3 helpers. Which items did they buy?

   ____

2. On Day 2, the Sports Club spent $14 and bought 4 items for 4 helpers. Which items did they buy?

   ____

3. On Day 3, the Sports Club spent $13 and bought 5 items for 5 helpers. Which items did they buy?

   ____

4. On Day 4, the Sports Club spent $21 and bought 6 items for 6 helpers. Which items did they buy?

   ____

Hint: Use your calculator or set up a spreadsheet to complete this table of prices of items from the Coffee House.

	Sausage Rolls	Cheese Skewers	Fruit Skewers	Muffins	Ham Roll-ups
Price	$2·50	$3·10	$2·75	$3·30	$3·90
1	$2·50	$3·10	$2·75	$3·30	$3·90
2					$7·80
3					$11·70
4					$15·60

Number & Algebra

AC9M5A01

Multiplication and Division

6 × 4 = 24, 4 × 6 = 24, 24 ÷ 6 = 4, 24 ÷ 4 = 6

These are all related facts.

Although 6 × 4 has the same answer as 4 × 6, they are not interchangeable when you are matching a number sentence to a problem.

How many wheels are there on 4 tricycles?

For this problem, you have to work out 4 lots of 3 wheels, so the number sentence should be written as 4 × 3.

Draw lines to match these word problems to the correct number sentences.

1. 6 quad bikes, how many wheels?
2. 24 oranges shared equally between 6 nets. How many in each net?
3. 24 oranges with 4 oranges in each net. How many nets?
4. 4 nets of oranges with 6 in each net. How many oranges altogether?

6 × 4 = 24
4 × 6 = 24
24 ÷ 6 = 4
24 ÷ 4 = 6

Write the correct number sentence for each of these word problems.

5. 4 trays of sausage rolls with 8 on each tray. How many sausage rolls altogether?

   ____

6. 32 sausage rolls with 8 on each tray. How many trays of sausage rolls are there?

   ____

7. 8 cars in the tyre shop. How many tyres to change? ____________
8. 32 sausage rolls shared equally between 4 trays. How many sausage rolls on each tray?

   ____

SCORE /8 0-3 4-6 7-8

Quick Check

Write 4 related number sentences for these numbers.

1. 15: ___ × ___ = ___, ___ × ___ = ___, ___ ÷ ___ = ___, ___ ÷ ___ = ___
2. 21: ___ × ___ = ___, ___ × ___ = ___, ___ ÷ ___ = ___, ___ ÷ ___ = ___
3. 42: ___ × ___ = ___, ___ × ___ = ___, ___ ÷ ___ = ___, ___ ÷ ___ = ___
4. 54: ___ × ___ = ___, ___ × ___ = ___, ___ ÷ ___ = ___, ___ ÷ ___ = ___
5. 56: ___ × ___ = ___, ___ × ___ = ___, ___ ÷ ___ = ___, ___ ÷ ___ = ___
6. 27: ___ × ___ = ___, ___ × ___ = ___, ___ ÷ ___ = ___, ___ ÷ ___ = ___
7. 28: ___ × ___ = ___, ___ × ___ = ___, ___ ÷ ___ = ___, ___ ÷ ___ = ___
8. 63: ___ × ___ = ___, ___ × ___ = ___, ___ ÷ ___ = ___, ___ ÷ ___ = ___
9. 33: ___ × ___ = ___, ___ × ___ = ___, ___ ÷ ___ = ___, ___ ÷ ___ = ___
10. 93: ___ × ___ = ___, ___ × ___ = ___, ___ ÷ ___ = ___, ___ ÷ ___ = ___

SCORE /10 0-4 5-8 9-10

TARGETING MATHS HOMEWORK: YEAR 5 © PASCAL PRESS ISBN: 9781925726589

Number & Algebra

UNIT 30

AC9M5N09

Multiplication of Money Amounts

The area model for multiplication can be used for money amounts. For example, to find the cost of 6 mangoes at $4·45 each, we can split the cents from the dollars and then split the ten cents from the single cents, like this:

Remember to convert the cents into dollars: 270c = $2·70.

	Dollars	Cents	
	4	40	5
× 6	24	240	30

= $26·70

Complete these money multiplications.

1. $3·65 × 8

	Dollars	Cents	
× ___			

= $___·___

2. $2·55 × 7

	Dollars	Cents	
× ___			

= $___·___

3. $4·63 × 5

	Dollars	Cents	
× ___			

= $___·___

4. $4·89 × 3

	Dollars	Cents	
× ___			

= $___·___

5. $3·35 × 4

	Dollars	Cents	
× ___			

= $___·___

6. $6·72 × 8

	Dollars	Cents	
× ___			

= $___·___

SCORE /6 0-2 3-4 5-6

Statistics & Probability

AC9M5P01

In the Bag

Ella put 12 counters into a bag for a chance experiment. There were red, green and blue counters in the bag.

Ella said, "There is a 1 in 6 chance of pulling a red counter from the bag and a 1 in 4 chance of pulling a green counter from the bag."

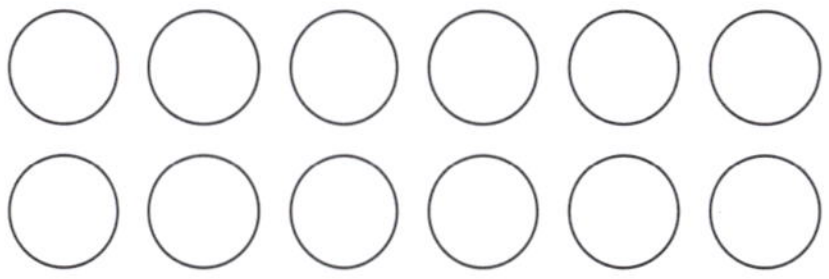

1. Colour the diagram to show how many counters of each colour there are.
2. What is the chance of pulling a blue counter from the bag?

   ___ in ____ chance

3. How many green counters need to be added to the bag so that there is a 50% chance of pulling a blue counter?

   ____ counters

4. What would then be the chance of pulling a red counter?

   ___ in ___ chance

SCORE /4 0-1 2-3 4

TERM 4

AC9M5SP01

3D Nets

None of these nets will fold to make a 3D object. Annotate the diagrams to show how to correct them.

1. 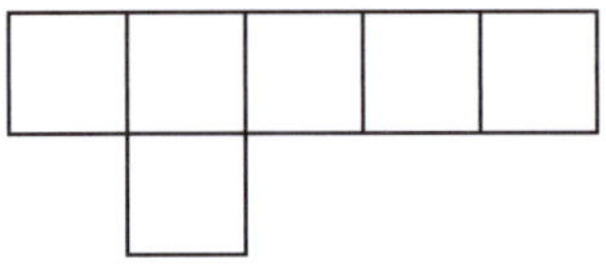

2. 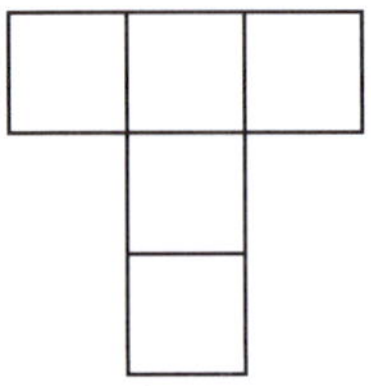

3. 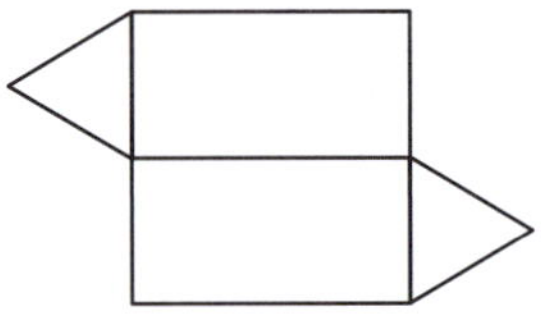

4. 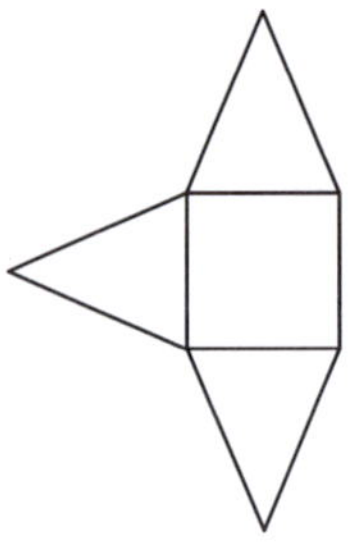

5. 

6. 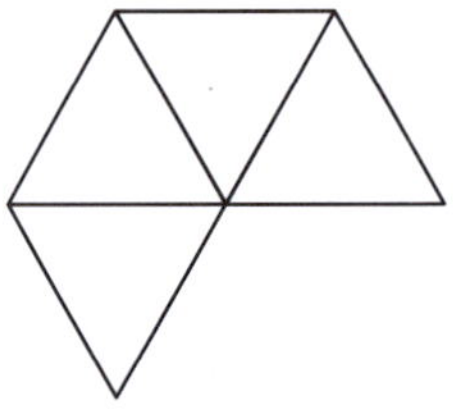

SCORE /6 (0-2) (3-4) (5-6)

NAPLAN*-style Questions

1. The greengrocer used this number sentence to solve one of his problems: 48 ÷ 8 = 6. Which word problem matches his number sentence?

   - ◯ The greengrocer has 48 oranges to put equally into 4 nets. How many nets will he need?
   - ◯ The greengrocer has 8 oranges to put equally into 48 nets. How many oranges should he put into each net?
   - ◯ The greengrocer has to share his 48 oranges equally between 8 nets. How many oranges should he put into each net?
   - ◯ The greengrocer has 6 oranges. He has to put the oranges equally into 42 nets. How many oranges should he put into each net?

2. These are the counters in a bag. What is the chance of not pulling out a yellow one?

   The chance is __ in __.

3. What is the shape whose net has six quadrilateral faces, with two different to the rest?

   - ◯ Square-based pyramid
   - ◯ Cube
   - ◯ Rectangular prism
   - ◯ Triangular prism

SCORE /3 (0-1) (2) (3)

TERM 4

TARGETING MATHS HOMEWORK: YEAR 5 © PASCAL PRESS ISBN: 9781925726589

Problem of the Week

UNIT 30

AC9M5SP03

Shape Puzzle

I have just been given a puzzle kit which has these shapes in it:

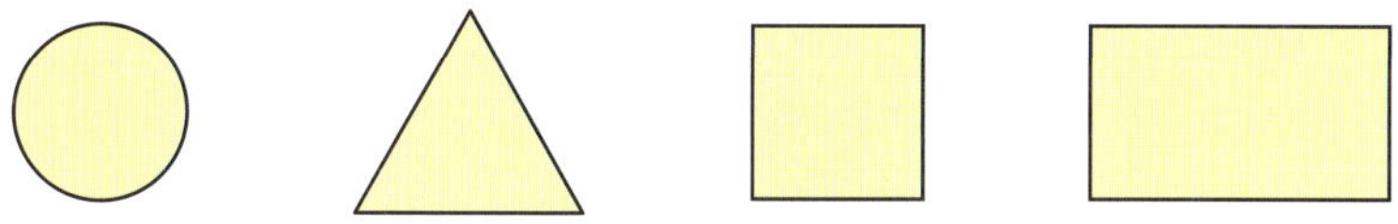

2 circles | 6 triangles | 8 squares | 8 rectangles

The instructions say that I can make five 3D objects with these pieces. Suggestions please!

Use the space below to create nets that use all of the puzzle pieces and discover which 3D objects Will could make with his kit.

Extra Challenge

Suppose Will had made a square-based pyramid before making any other 3D objects.

Draw the 2D pieces he would have left over that could not be used to make a 3D object.

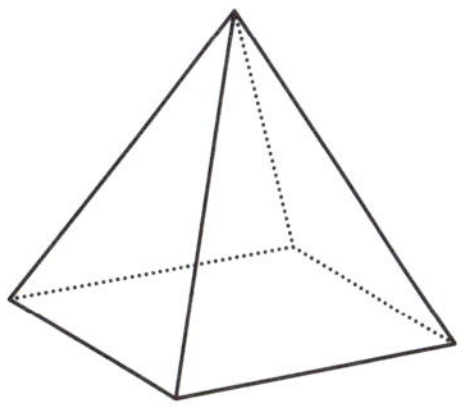

TERM 4

Number & Algebra

AC9M5N04

Percentages

These shapes are drawn on 100-square grids. What percentage of each grid is coloured?

① _____%

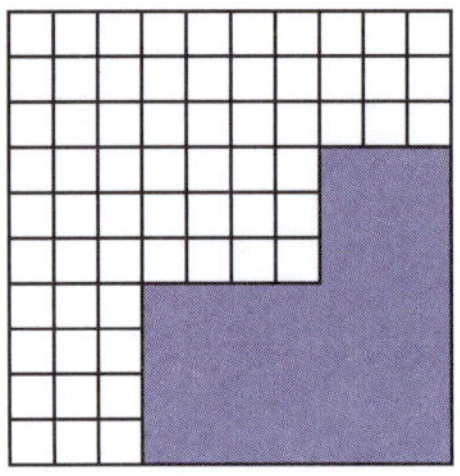

② _____%

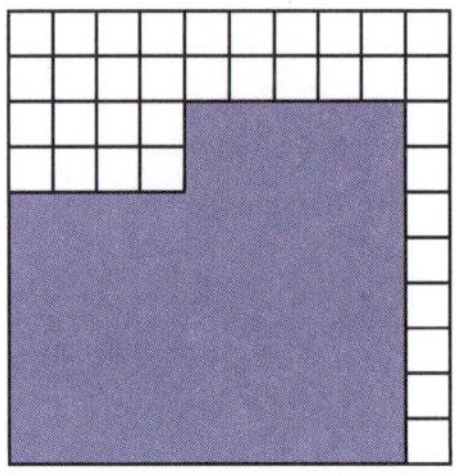

③ _____%

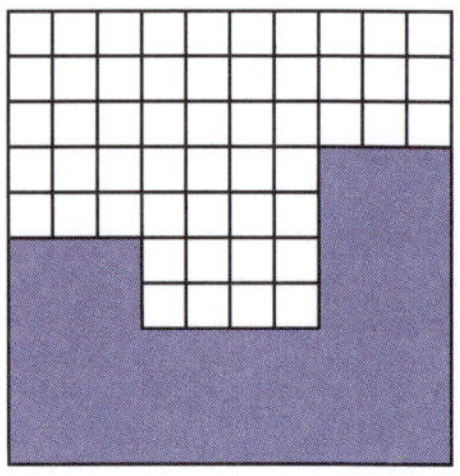

What percentage of each grid is uncoloured?

④ _____%

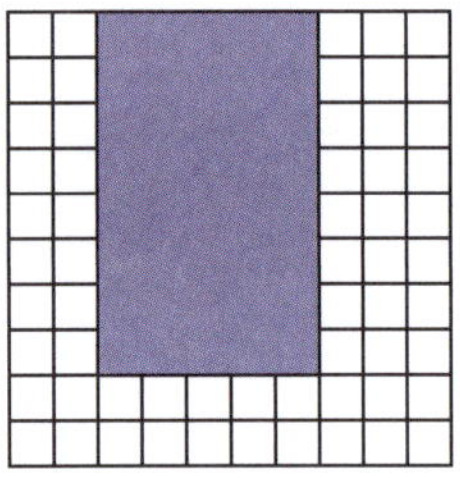

⑤ _____%

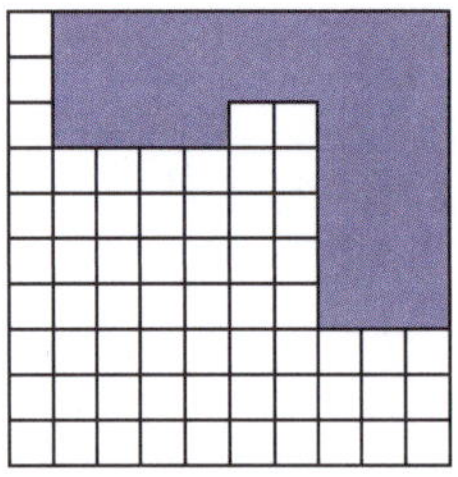

⑥ _____%

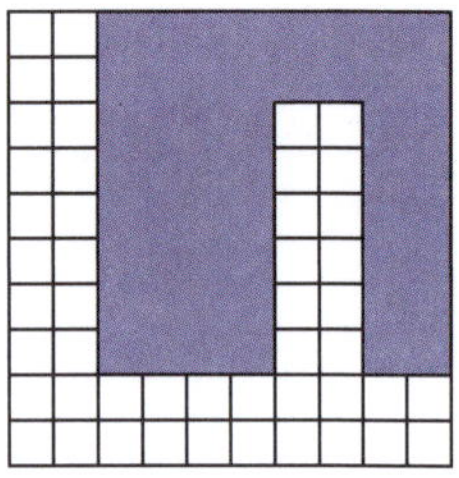

Write the percentage of the grid covered by each colour.

⑦

Yellow _____%
Green _____%
Blue _____%

⑧

Yellow _____%
Green _____%
Blue _____%

⑨

Yellow _____%
Green _____%
Blue _____%

SCORE /9 0-4 5-7 8-9

Quick Check

Complete these percentage tables.

	Coloured	Uncoloured
①	70%	
②	35%	
③		85%
④		63%
⑤	24%	

	Red	Orange	Purple
⑥	40%	26%	
⑦	30%		27%
⑧		42%	16%
⑨		19%	28%
⑩	3%	12%	

SCORE /10 0-4 5-8 9-10

TARGETING MATHS HOMEWORK: YEAR 5 © PASCAL PRESS ISBN: 9781925726589

Number & Algebra

UNIT 31

AC9M5N03

Fraction Word Problems

Fractions have many everyday applications.
Solve these everyday fraction problems.

1. It takes $\frac{1}{2}$ hour to walk the dog and $\frac{3}{4}$ hour to make dinner.
   How long is that altogether?
   ____

2. Asha had $3\frac{1}{2}$ metres of fabric.
   She used $1\frac{3}{4}$ metres to make a shirt.
   How much fabric does Asha have left?
   ____

3. There was $1\frac{3}{4}$ kilograms of flour in the container.
   Ravi used $\frac{1}{2}$ kilogram to make muffins.
   How much flour was left then?
   ____

4. The filtered water tower had $5\frac{1}{2}$ litres of water.
   Sam filled three $\frac{3}{4}$ litre water bottles from it.
   How much water was left in the water tower then?
   ____

The Naidoo family are travelling to Adelaide for their holiday. Adelaide is 500 kilometres away. They plan to stop for morning tea $\frac{1}{10}$ of the way through the journey and then for a snack $\frac{3}{4}$ of the way through the journey.

5. How far will they have travelled by morning tea? ______________
6. How far will they have travelled when they stop for a snack? ______________

SCORE /6 0-2 3-4 5-6

Statistics & Probability

AC9M5P02

On the Road

The Kang family are driving from Adelaide to the Gold Coast which is 2022 kilometres away.

On a scale of 0 to 10, with 0 being the lowest chance (impossible) and 10 being the highest chance (certain), what do you think are the chances of these events occurring?

1. They will have a puncture. ___
2. They will stay overnight somewhere. ___
3. They will get lost. ___
4. They will be pleased to get to the Gold Coast. ___
5. They will be looking forward to going home. ___

SCORE /5 0-2 3-4 5

TERM 4

AC9M5M04

Angles

"Good morning class. Today I want you to draw a zig-zag line with 5 angles above the line and 5 angles below the line, just like this one."

"Then I want you to:
- measure the total length of your line
- measure the angles above your line and find their total
- measure the angles below your line and find their total."

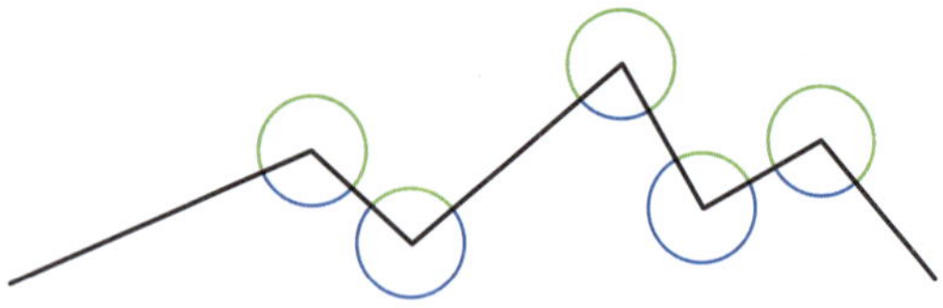

Draw a zig-zag line with 5 angles like Mrs Green's.

1. Use your ruler.
   What is the total length of your line? ________
2. Use your protractor to measure the angles above the line and below the line.
   Complete the table.
3. The angle above the line and below the line always add to _____°.

Angle	Above	Below	Total
Angle 1	°	°	°
Angle 2	°	°	°
Angle 3	°	°	°
Angle 4	°	°	°
Angle 5	°	°	°

SCORE /7 0-3 4-5 6-7

NAPLAN*-style Questions

1. Ravi has paid 57% of the $100 for the school trip.
   How much does Ravi still owe for the trip?
   ◯ $100 ◯ $43 ◯ $53 ◯ $157
2. This quadrilateral has four different angles.
   Which angle is closest to 45°?
   ◯ A ◯ B ◯ C ◯ D

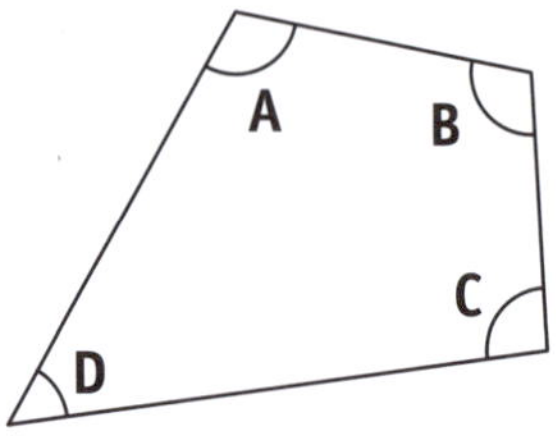

3. We cut five oranges into quarters to make slices for the football teams.
   How many slices did we make?
   ◯ 4 ◯ 5 ◯ 13 ◯ 20

SCORE /3 0-1 2 3

TARGETING MATHS HOMEWORK: YEAR 5 © PASCAL PRESS ISBN: 9781925726589

Problem of the Week

AC9M5A02

The Calculator Kids

The Calculator Kids are making number sequences on their calculators.

I entered 4 + 3 = = = ...

I entered 2 + 4 = = = ...

Do you think the two sequences will ever show the same number in the display? Enter their key sequences on your calculator to find out. Record your results.

1. 4 + 3 = = = ... makes the sequence:

___ ___ ___ ___ ___ ___ ___ ___ ___

2. 2 + 4 = = = ... makes the sequence:

___ ___ ___ ___ ___ ___ ___ ___ ___

Then the Calculator Kids had another go.

I entered 3 + 4 = = = ...

I entered 2 + 6 = = = ...

Do you think the two sequences will ever show the same number in the display? Enter their key sequences on your calculator to find out. Record your results.

3. 3 + 4 = = = ... makes the sequence:

___ ___ ___ ___ ___ ___ ___ ___ ___

4. 2 + 6 = = = ... makes the sequence:

___ ___ ___ ___ ___ ___ ___ ___ ___

5. What happened this time? ______________________

6. Can you see why? ______________________

The Challenge

7. Find some other starting numbers that **do** give the same number in the display.

____

8. Find some other starting numbers that will **never** give the same numbers in the display.

____

Working space

AC9M5A02

Finding Unknowns

The questions below ask you to find an unknown, which is a missing number in a number sentence. Here is an example:

6 × ? + 3 = 21

To remove the 3 from the first part, subtract 3 from both sides. Then solve the problem of finding the unknown.

6 × ? + 3 **– 3** = 21 **– 3**

6 × ? = 18

6 × **3** = 18

Complete the steps that Miro would use to find these unknowns.

1. 7 × ? + 5 = 61
   7 × ? + 5 – _____ = 61 – _____
   7 × ? = _____
   7 × _____ = _____

2. 4 × ? + 3 = 67
   4 × ? + 3 – _____ = 67 – _____
   4 × ? = _____
   4 × _____ = _____

3. 6 × ? – 5 = 37
   6 × ? – 5 + _____ = 37 + _____
   6 × ? = _____
   6 × _____ = _____

4. ? × 6 + 3 = 75
   ? × 6 + 3 – _____ = 75 – _____
   ? × 6 = _____
   _____ × 6 = _____

5. ? × 5 + 6 = 126
   ? × 5 + 6 – _____ = 126 – _____
   ? × 5 = _____
   _____ × 5 = _____

6. ? × 8 – 5 = 51
   ? × 8 – 5 + _____ = 51 + _____
   ? × 8 = _____
   _____ × 8 = _____

SCORE /6 0-2 3-4 5-6

TERM 4

Quick Check

Complete these multiplications and divisions.

1. 6 × ____ = 72
2. ____ × 23 = 69
3. 7 × ____ = 350
4. ____ × 25 = 200
5. 9 × ____ = 189
6. 64 ÷ ____ = 8
7. 39 ÷ ____ = 13
8. ____ ÷ 12 = 5
9. ____ ÷ 25 = 7
10. ____ ÷ 11 = 20

SCORE /10 0-4 5-8 9-10

TARGETING MATHS HOMEWORK: YEAR 5 © PASCAL PRESS ISBN: 9781925726589

Number & Algebra

AC9M5A02

Unknowns in Money Problems

The strategy we used to find unknowns in number sentences can also be used to find unknowns in money problems. Here is an example:

What does a pair of shoes cost if three pairs cost $45?

This is like asking: **3 at how much makes $45?**

As before, write a sentence with the unknown.
Divide both sides by 3 to find the unknown: $? × 3 = $45

$? = $45 **÷ 3**

$? = **$15**

Complete the steps in the strategy that Ella would use to find these unknowns.

1. For 6 at $[ ] each = $72 use:
   $? × _____ = $72
   $? = $72 ÷ _____
   $? = _________
2. For 8 at $[ ] each = $96 use:
   $? × 8 = $_____
   $? = $_____ ÷ _____
   $? = _________
3. For 3 at $[ ] each = $18·15 use:
   $? × 3 = $_______
   $? = $_______ ÷ _____
   $? = _________
4. For 7 at $[ ] each = $84·70 use:
   $? × _____ = $_______
   $? = $_______ ÷ _____
   $? = _________
5. For 4 at $[ ] each = $4·44 use:
   $? × _____ = $_______
   $? = $_______ ÷ _____
   $? = _________
6. For 9 at $[ ] each = $108·90 use:
   $? × _____ = $_______
   $? = $_______ ÷ _____
   $? = _________

SCORE /6 0-2 3-4 5-6

Statistics & Probability

AC9M5ST03

Family Movie Night

The parents and friends committee is organising a family night out at the movies but they can't decide which two movies to show. They thought they should do a survey to make sure that both parents and students will enjoy the movies.

Write five questions that you think the committee should include in their survey.

1. ______________________________
2. ______________________________
3. ______________________________
4. ______________________________
5. ______________________________

SCORE /5 0-2 3-4 5

TERM 4

Measurement & Space

AC9M5SP02

Hidden Picture

Follow the instructions below to find the hidden picture in this grid.

This is like a join-the-dots picture except that you have to position the dots on the grid before you can join them.

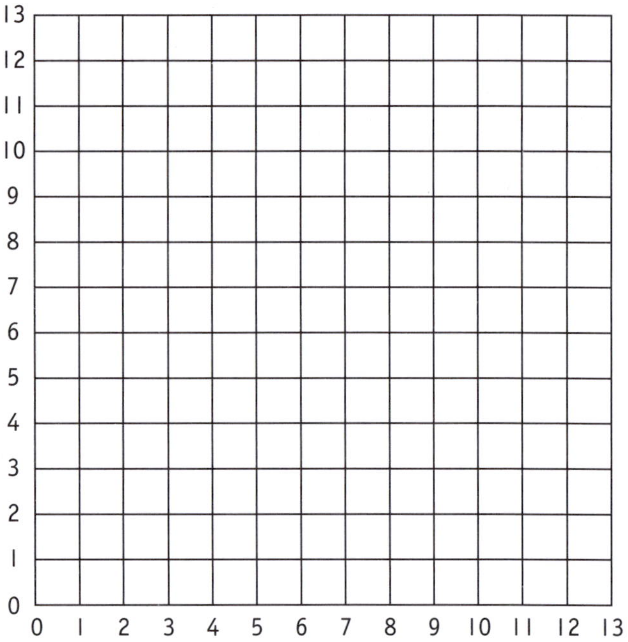

For each list of points, first make a dot at each point on the grid.

Then start at the first point in the list and join each point to the next point in the list.

1. List 1. Mark dots at (5, 1), (8, 1), (11, 3), (11, 9), (2, 9), (2, 3), (5, 1) and join the dots in that order.
2. List 2. Mark dots at (2, 8), (1, 11), (4, 9) and join the dots in that order.
3. List 3. Mark dots at (9, 9), (12, 11), (11, 8) and join the dots in that order.
4. List 4. Mark dots at (5, 2), (8, 2), (8, 3), (5, 3), (5, 2) and join the dots in that order.
5. List 5. Mark dots at (3, 7), (5, 7), (5, 6), (3, 6), (3, 7) and join the dots in that order.
6. List 6. Mark dots at (8, 7), (10, 7), (10, 6), (8, 6), (8, 7) and join the dots in that order.
7. List 7. Mark dots at (6, 4), (7, 4), (7, 7), (6, 7), (6, 4) and join the dots in that order.

Embellish your picture with colour and extra detail.

SCORE /7 0-3 4-5 6-7

NAPLAN*-style Questions

1. Asha bought 5 identical notebooks and spent $37 altogether. How much was each notebook?

   $ ______

2. Which set of coordinates matches the × on this grid?

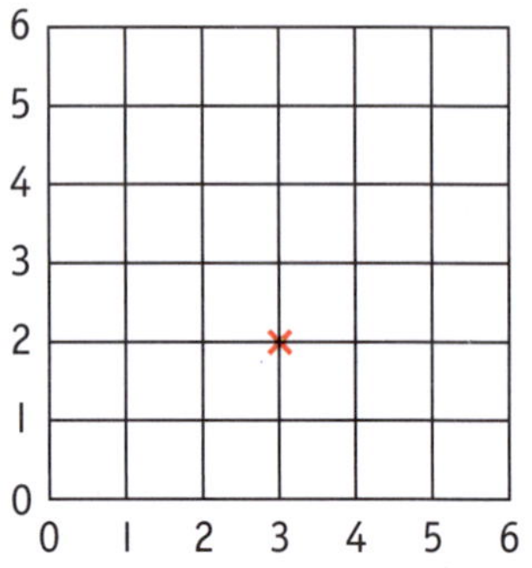

   ◯ (2, 3) ◯ (3, 3) ◯ (3, 2) ◯ (4, 2)

3. Complete this number sentence.

   8 × ☐ = 128

SCORE /3 0-1 2 3

TARGETING MATHS HOMEWORK: YEAR 5 © PASCAL PRESS ISBN: 9781925726589

AC9M5N08

The Fundraiser

The Sports Club are planning a fundraising sausage sizzle.
They hope to sell 120 sausages in bread rolls at $3·60 each.

- A tray of 8 sausages costs $7·95.
- A bag of 12 bread rolls costs $10·40.

① How many trays of sausages and bags of bread rolls will they need?

② If they sell all the sausages in bread rolls, how much money will they collect?

③ When they deduct the cost of the sausages and bread rolls, how much profit will they be left with?

④ What might the Sports Club do to raise more money?

Review: Unit 25 and Unit 26

AC9M5N01, AC9M5N03, AC9M5N04, AC9M5N07, AC9M5M04, AC9M5ST02, AC9M5P02

Write these fractions in expanded decimal form.

1. $\frac{11}{1000}$ is ___ ones, ___ tenths, ___ hundredths and ___ thousandths.
2. $\frac{109}{1000}$ is ___ ones, ___ tenths, ___ hundredths and ___ thousandths.
3. $\frac{19}{10}$ is ___ ones, ___ tenths, ___ hundredths and ___ thousandths.

Answer the questions about this travel line graph.

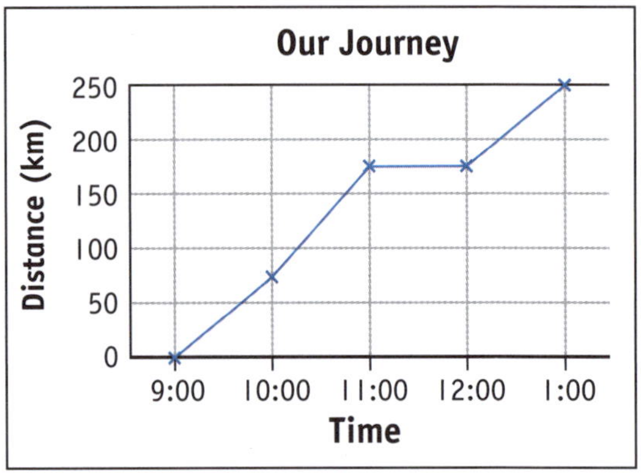

4. At what time did our journey begin?

   ____

5. How far did we travel?

   ____

6. Between which two times was the fastest part of our journey?

   ___________ and ___________

7. At what time did we stop for a break?

   ____

8. How large are these angles? Use your protractor to check.

   Angle A is _____°.

   Angle B is _____°.

   Angle C is _____°.

   Angle D is _____°.

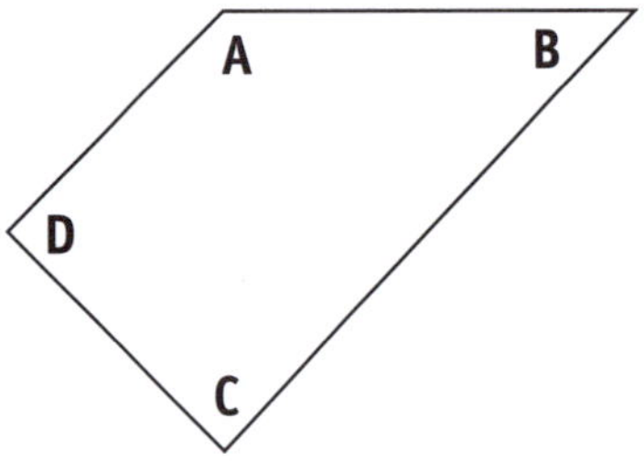

Change the three fractions in each group into equivalent fractions with the same denominator.

9. $1\frac{7}{8} = \frac{}{}$ $\frac{3}{4} = \frac{}{}$ $1\frac{1}{2} = \frac{}{}$
10. $\frac{2}{5} = \frac{}{}$ $\frac{1}{2} = \frac{}{}$ $1\frac{3}{10} = \frac{}{}$

Work out the percentages of these amounts.

11. 50% of $19 = $_______
12. 50% of $105 = $_______
13. 25% of $50 = $_______
14. 75% of $50 = $_______

15. Which spinner has a 25% chance of spinning blue?

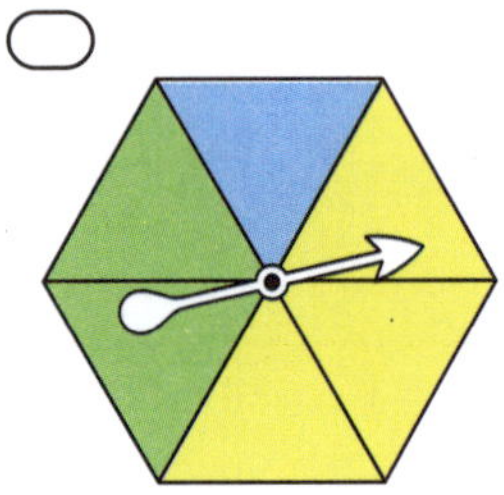
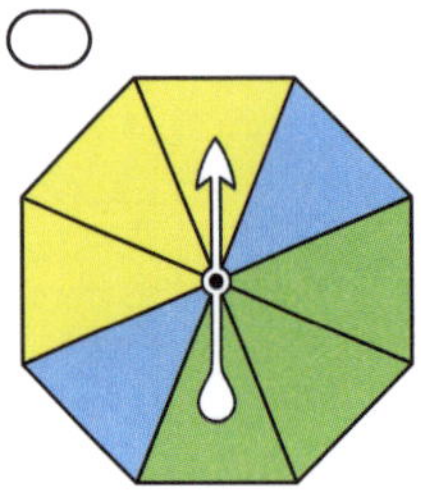
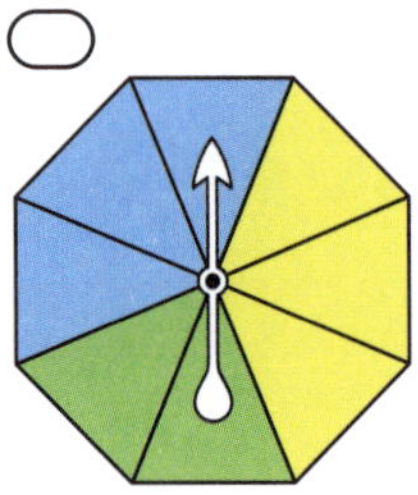

SCORE /15 0-7 8-13 14-15

TARGETING MATHS HOMEWORK: YEAR 5 © PASCAL PRESS ISBN: 9781925726589

Review: Unit 27 and Unit 28

AC9M5N05, AC9M5N06, AC9M5N10, AC9M5ST01

Use colour to show compatible fractions and then write the answers.

1. $\frac{3}{4} + \frac{1}{5} + \frac{2}{8} =$
2. $\frac{3}{5} + \frac{1}{2} + \frac{4}{10} =$
3. $\frac{1}{3} + \frac{4}{6} + \frac{3}{5} =$
4. $\frac{5}{8} + \frac{1}{4} + \frac{1}{6} + \frac{1}{8} =$

Use the commutative property to multiply these numbers.

5. 3 × 4 × 5 × 6 ____ × ____ = ____, ____ × ____ = ____, ____ × ____ = ____
6. 7 × 2 × 6 × 5 ____ × ____ = ____, ____ × ____ = ____, ____ × ____ = ____

Use the graph of borrowed library books to answer the questions.

7. What is the mode of the data? ______
8. How many Picture and Joke books were borrowed altogether?

   ______
9. How many fewer Joke books were borrowed than Fiction books?

   ______

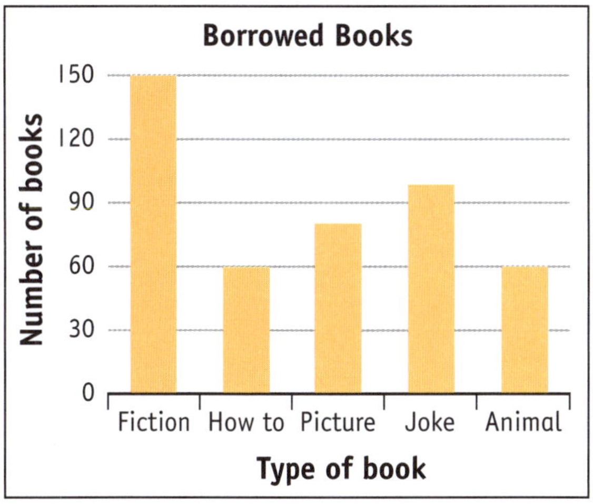

Use factors and rules of divisibility to answer these questions.

10. 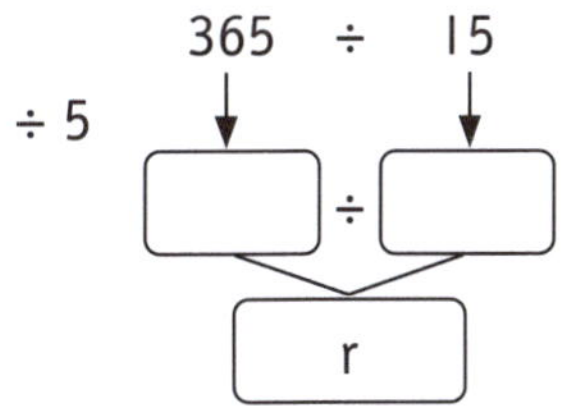

11. 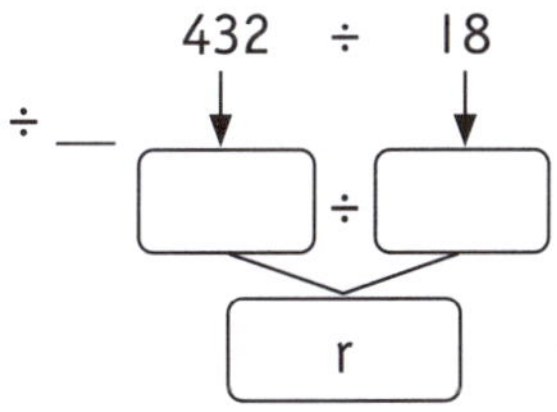

12. 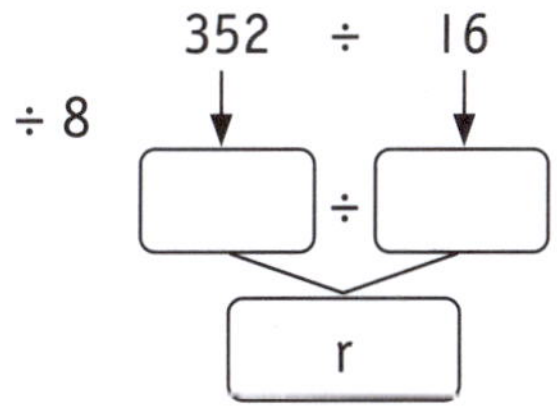

Use factors for these multiplications.

13. 18 × 16

    = ___ × ___ × ___ × ___

    ___ × ___ × ___ × ___ = ____

14. 24 × 15

    = ___ × ___ × ___ × ___

    ___ × ___ × ___ × ___ = ____

Use the shapes on the grid to answer the following questions.

15. What are the colours of the two shapes that have the same perimeter?

    ________ and ________
16. What are the colours of the two shapes that have the same area?

    ________ and ________

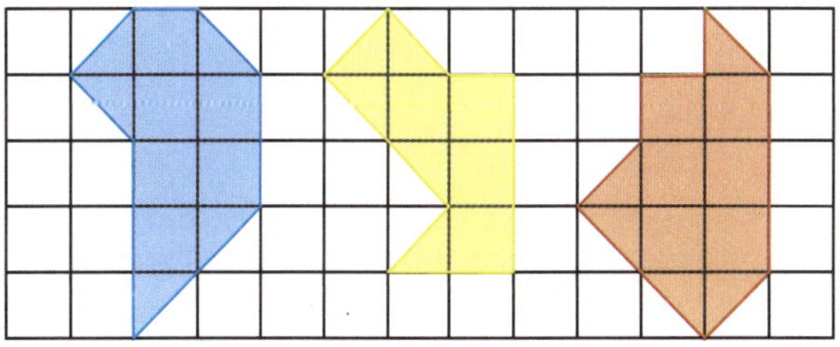

SCORE /16 0-7 8-14 15-16

Review: Unit 29 and Unit 30

AC9M5N02, AC9M5N08, AC9M5N09, AC9M5A01, AC9M5M03

(1) Fill in the missing values in the School Order spreadsheet.

	Pencils	Pens	Notebooks	Erasers
1	$1·35	$1·95	$2·60	$0·89
2				
3				
4				

(2) How much for 10 pencils? $________

(3) How much for 20 pens? $________

(4) Find the lowest common multiple of 9 and 12.

9 12

___ × ___ × ___ = ____

(5) Sydney is 11 hours ahead of London. When it is 6 am in London, the time in Sydney is ________ on the same day / next day / previous day.

(6) There are six bananas on each bunch and eight bunches altogether.
How many bananas are there?

____ bananas

(7) If there are 72 bananas on 8 bunches that are all the same size,
how many bananas are there on each bunch?

____ bananas

(8) If there are 49 bananas with 7 bananas on each bunch,
how many bunches of bananas are there?

____ bunches

Complete these money multiplications.

(9) $2·65 × 6

Dollars	Cents	

× __ = $___·___

(10) $3·36 × 7

Dollars	Cents	

× __ = $___·___

(11) $8·43 × 3

Dollars	Cents	

× __ = $___·___

(12) $6·45 × 8

Dollars	Cents	

× __ = $___·___

SCORE /12 0-5 6-10 11-12

TARGETING MATHS HOMEWORK: YEAR 5 © PASCAL PRESS ISBN: 9781925726589

Review: Unit 31 and Unit 32

AC9M5N03, AC9M5N04, AC9M5A02, AC9M5M04, AC9M5SP02

Complete these percentage tables.

	Coloured	Uncoloured
1	37%	____
2	24%	____
3	____	35%

	Red	Green	Blue
4	30%	27%	____
5	28%	____	56%
6	____	45%	16%

7. Ravi spent $\frac{1}{4}$ hour mixing a cake and $\frac{3}{4}$ hour cooking it.
The cake then took $\frac{1}{6}$ hour to cool and $\frac{1}{3}$ hour to decorate.
How long did it take Ravi to make the cake? _____ hours

8. How long is the zig-zag line?
_____ mm

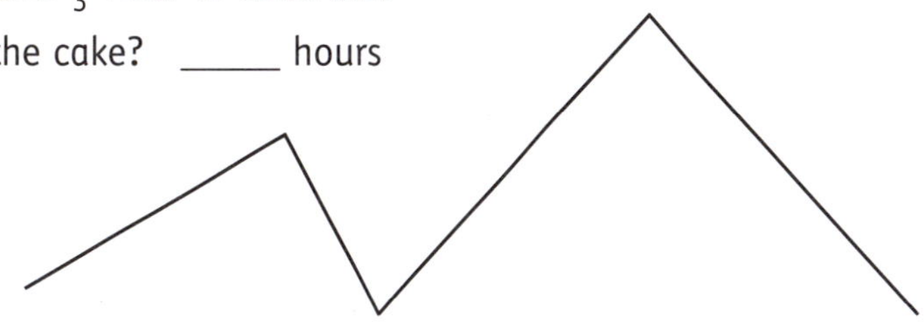

Estimate and then measure the size of these angles.

9.

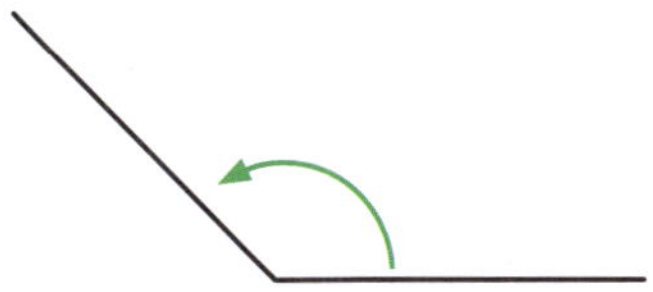

Estimate: _____°
Measure: _____°

10.

Estimate: _____°
Measure: _____°

11.

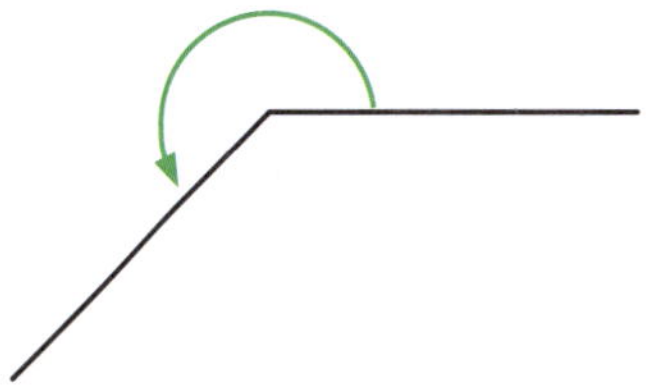

Estimate: _____°
Measure: _____°

Work out these unknowns.

12. 8 × ? + 6 = 54
8 × ? + 6 − ____ = 54 − ____
8 × ? = ____
8 × ____ =

13. ? × 6 + 5 = 47
? × 6 + 5 − ____ = 47 − ____
? × 6 − ____
____ × 6 = ____

Use related division facts to answer these questions.

14. For 6 at $☐ each = $96 use:
$? × 6 = $____
$? = $____ ÷ ____
$? = $____

15. For 8 at $☐ each = $96·80 use:
$? × 8 = $______
$? = $______ ÷ ____
$? = $______

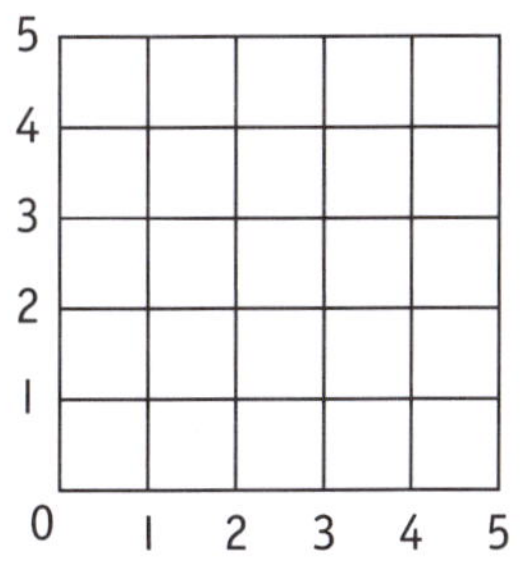

16. Mark dots at (1, 2), (4, 5), (5, 3), (1, 1) and (1, 2) and join the dots in that order.

17. How many right angles have you created? ____

18. How many squares have you enclosed? ____

19. What is the name of a shape with four sides? ____________________

SCORE /19 0-9 10-17 18-19

TARGETING MATHS HOMEWORK: YEAR 5 © PASCAL PRESS ISBN: 9781925726589

★ QUICK MATHS FACTS ★

Measurement

Length

1 km = 1000 m

1 m = 100 cm

1 cm = 10 mm

Area

1 m^2 = 10 000 cm^2

1 hectare = 10 000 m^2

1 km^2 = 1 000 000 m^2

Capacity

1 L = 1000 mL

Mass

1 t = 1000 kg

1 kg = 1000 g

Fractions, decimals and percentages

$\frac{1}{100}$ = 0·01 = 1%

$\frac{25}{100}$ = 0·25 = 25%

$\frac{50}{100}$ = 0·5 = 50%

$\frac{75}{100}$ = 0·75 = 75%

$\frac{100}{100}$ = 1 = 100%

Place value

Ones	1
Tens	12
Hundreds	123
Thousands	1234
Ten thousands	12 345
Hundred thousands	123 456
Millions	1 234 567
Ten millions	12 345 678
Hundred millions	123 456 789

Time

60 s = 1 min

60 min = 1 h

1440 min = 1 day

24 h = 1 day

7 days = 1 week

365 days = 1 year

52 weeks = 1 year

12 months = 1 year

Multiplication tables

1 × 1 = 1	2 × 1 = 2	3 × 1 = 3	4 × 1 = 4	5 × 1 = 5	6 × 1 = 6
1 × 2 = 2	2 × 2 = 4	3 × 2 = 6	4 × 2 = 8	5 × 2 = 10	6 × 2 = 12
1 × 3 = 3	2 × 3 = 6	3 × 3 = 9	4 × 3 = 12	5 × 3 = 15	6 × 3 = 18
1 × 4 = 4	2 × 4 = 8	3 × 4 = 12	4 × 4 = 16	5 × 4 = 20	6 × 4 = 24
1 × 5 = 5	2 × 5 = 10	3 × 5 = 15	4 × 5 = 20	5 × 5 = 25	6 × 5 = 30
1 × 6 = 6	2 × 6 = 12	3 × 6 = 18	4 × 6 = 24	5 × 6 = 30	6 × 6 = 36
1 × 7 = 7	2 × 7 = 14	3 × 7 = 21	4 × 7 = 28	5 × 7 = 35	6 × 7 = 42
1 × 8 = 8	2 × 8 = 16	3 × 8 = 24	4 × 8 = 32	5 × 8 = 40	6 × 8 = 48
1 × 9 = 9	2 × 9 = 18	3 × 9 = 27	4 × 9 = 36	5 × 9 = 45	6 × 9 = 54
1 × 10 = 10	2 × 10 = 20	3 × 10 = 30	4 × 10 = 40	5 × 10 = 50	6 × 10 = 60
1 × 11 = 11	2 × 11 = 22	3 × 11 = 33	4 × 11 = 44	5 × 11 = 55	6 × 11 = 66
1 × 12 = 12	2 × 12 = 24	3 × 12 = 36	4 × 12 = 48	5 × 12 = 60	6 × 12 = 72

7 × 1 = 7	8 × 1 = 8	9 × 1 = 9	10 × 1 = 10	11 × 1 = 11	12 × 1 = 12
7 × 2 = 14	8 × 2 = 16	9 × 2 = 18	10 × 2 = 20	11 × 2 = 22	12 × 2 = 24
7 × 3 = 21	8 × 3 = 24	9 × 3 = 27	10 × 3 = 30	11 × 3 = 33	12 × 3 = 36
7 × 4 = 28	8 × 4 = 32	9 × 4 = 36	10 × 4 = 40	11 × 4 = 44	12 × 4 = 48
7 × 5 = 35	8 × 5 = 40	9 × 5 = 45	10 × 5 = 50	11 × 5 = 55	12 × 5 = 60
7 × 6 = 42	8 × 6 = 48	9 × 6 = 54	10 × 6 = 60	11 × 6 = 66	12 × 6 = 72
7 × 7 = 49	8 × 7 = 56	9 × 7 = 63	10 × 7 = 70	11 × 7 = 77	12 × 7 = 84
7 × 8 = 56	8 × 8 = 64	9 × 8 = 72	10 × 8 = 80	11 × 8 = 88	12 × 8 = 96
7 × 9 = 63	8 × 9 = 72	9 × 9 = 81	10 × 9 = 90	11 × 9 = 99	12 × 9 = 108
7 × 10 = 70	8 × 10 = 80	9 × 10 = 90	10 × 10 = 100	11 × 10 = 110	12 × 10 = 120
7 × 11 = 77	8 × 11 = 88	9 × 11 = 99	10 × 11 = 110	11 × 11 = 121	12 × 11 = 132
7 × 12 = 84	8 × 12 = 96	9 × 12 = 108	10 × 12 = 120	11 × 12 = 132	12 × 12 = 144

Days in the month

Thirty days have September, April, June and November.
All the rest have thirty-one, except February alone...
which has twenty-eight days clear, and twenty-nine each leap year.

Answers - Homework Year 5

TERM 1

UNIT 1 NUMBER & ALGEBRA

Decimal Numbers and Algebra

1 0·01
2 0·19
3 0·008
4 0·23
5 0·013
6 0·143
7 1·2
8 1·2
9 1·2
10 0·156

Quick Check

1 $\frac{1}{10}$
2 $\frac{3}{100}$
3 $\frac{4}{1000}$
4 $\frac{9}{10}$
5 $\frac{29}{100}$
6 $\frac{235}{1000}$
7 $\frac{165}{1000}$
8 $\frac{16}{1000}$
9 $\frac{73}{100}$
10 $\frac{745}{1000}$

Renaming Decimals for Addition

1 $0·7 + 0·5 = \frac{7}{10} + \frac{5}{10} = \frac{12}{10}$ or 1·2
2 $0·9 + 0·3 = \frac{9}{10} + \frac{3}{10} = \frac{12}{10}$ or 1·2
3 $0·06 + 0·09 = \frac{6}{100} + \frac{9}{100} = \frac{15}{100}$ or 0·15
4 $0·07 + 0·08 = \frac{7}{100} + \frac{8}{100} = \frac{15}{100}$ or 0·15
5 $0·008 + 0·009 = \frac{8}{1000} + \frac{9}{1000} = \frac{17}{1000}$ or 0·017
6 $0·9 + 0·7 = \frac{9}{10} + \frac{7}{10} = \frac{16}{10}$ or 1·6
7 $0·006 + 0·005 = \frac{6}{1000} + \frac{5}{1000} = \frac{11}{1000}$ or 0·011
8 $0·07 + 0·06 = \frac{7}{100} + \frac{6}{100} = \frac{13}{100}$ or 0·13
9 $0·008 + 0·002 = \frac{8}{1000} + \frac{2}{1000} = \frac{10}{1000}$ or 0·01
10 $0·08 + 0·03 = \frac{8}{100} + \frac{3}{100} = \frac{11}{100}$ or 0·11

UNIT 1 STATISTICS & PROBABILITY

Describing the Chance

1 1 in 2 or $\frac{1}{2}$ chance
2 2 in 3 or $\frac{2}{3}$ chance
3 2 in 4 or $\frac{2}{4}$ chance (Allow $\frac{1}{2}$)
4 1 in 4 or $\frac{1}{4}$ chance
5 3 in 5 or $\frac{3}{5}$ chance

UNIT 1 MEASUREMENT & SPACE

Units of Measure

1 metre, m
2 gram, g
3 square centimetre, cm^2
4 millilitre, mL
5 kilogram, kg
6 centimetre, cm
7 litre, L
8 millimetre, mm
9 kilometre, km
10 centimetre, cm or metre, m
11 kilogram, kg
12 year, y

UNIT 1 NAPLAN*-STYLE QUESTIONS

1 0·01
2 Second spinner
3 0·5 kg

UNIT 1 PROBLEM OF THE WEEK

Just by Chance

Open-ended

UNIT 2 NUMBER & ALGEBRA

Factors

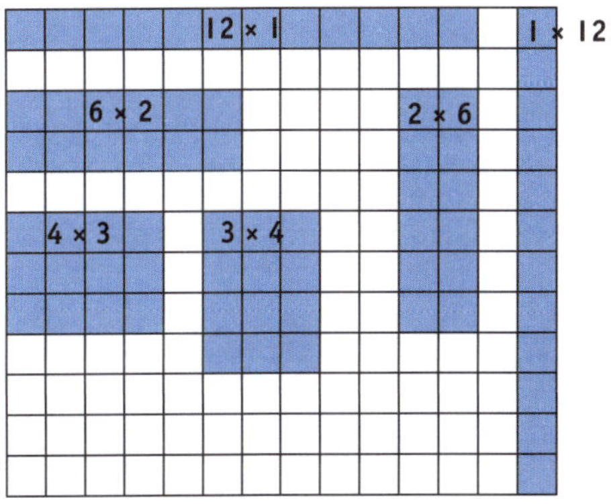

Quick Check

1 2: 1, 2
2 3: 1, 3
3 4: 1, 2, 4
4 7: 1, 7
5 8: 1, 2, 4, 8
6 9: 1, 3, 9
7 10: 1, 2, 5, 10
8 11: 1, 11
9 14: 1, 2, 7, 14
10 15: 1, 3, 5, 15

Multiples

	Number	2	3	4	5	6	7	8	9
1	12	✓	✓	✓		✓			
2	15		✓		✓				
3	20	✓		✓	✓				
4	24	✓	✓	✓		✓		✓	
5	28	✓		✓			✓		
6	30	✓	✓		✓	✓			
7	36	✓	✓	✓		✓			✓
8	42	✓	✓			✓	✓		
9	45		✓		✓				✓
10	56	✓		✓			✓	✓	

UNIT 2 STATISTICS & PROBABILITY

Show Ratings

	Answer	Table
1	21	Voters
2	1	Ratings
3	Sci-fi	Ratings
4	2–4	Either
5	Doco	Ratings

UNIT 2 MEASUREMENT & SPACE

Made with Shapes

1 Square-based pyramid
2 Rectangular prism
3

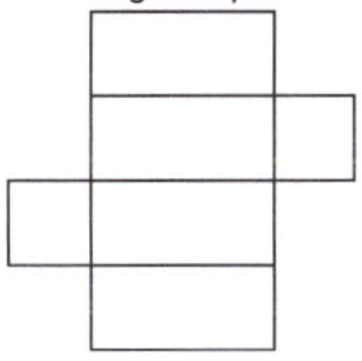

4

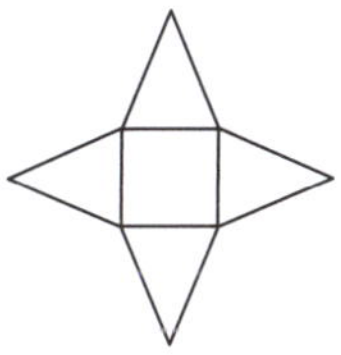

5 cylinder, cone

UNIT 2 NAPLAN*-STYLE QUESTIONS

1 12
2

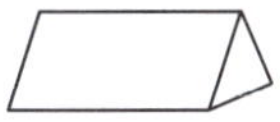

3 9

UNIT 2 PROBLEM OF THE WEEK

Shape Puzzle

Object 1: Ravi
Object 2: Miro
Object 3: Sam
Object 4: Asha
Object 5: Jade
Object 6: Leo

UNIT 3 NUMBER & ALGEBRA

Ordering Mixed Fractions

1 $1\frac{3}{5}$, $\frac{6}{5}$, $\frac{9}{5}$ $\frac{8}{5}$, $\frac{6}{5}$, $\frac{9}{5}$

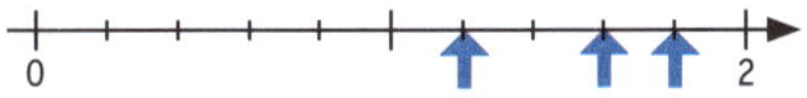

2 $\frac{1}{2}$, $1\frac{1}{4}$, $\frac{7}{4}$ $\frac{2}{4}$, $\frac{5}{4}$, $\frac{7}{4}$

ANSWERS

3 $\frac{1}{6}$, $1\frac{1}{6}$, $\frac{11}{6}$ $\frac{1}{6}$, $\frac{7}{6}$, $\frac{11}{6}$

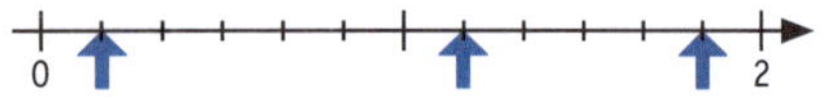

4 $\frac{12}{4}$, $1\frac{1}{4}$, $2\frac{3}{4}$ $\frac{12}{4}$, $\frac{5}{4}$, $\frac{11}{4}$

5 $\frac{5}{3}$, $1\frac{1}{3}$, $\frac{9}{3}$ $\frac{5}{3}$, $\frac{4}{3}$, $\frac{9}{3}$

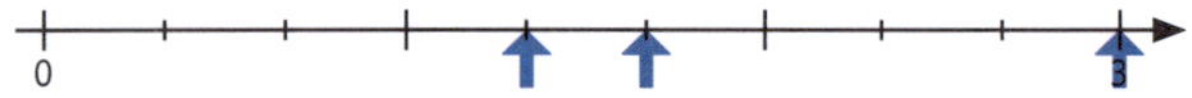

Quick Check

1 $\frac{11}{8}$
2 $\frac{7}{5}$
3 $\frac{15}{8}$
4 $1\frac{6}{10}$
5 $3\frac{1}{6}$
6 $\frac{15}{4}$
7 $\frac{9}{5}$
8 $\frac{13}{2}$
9 $\frac{16}{3}$
10 $\frac{15}{7}$

Fractions and Addition

1 $\frac{1}{6} + \frac{4}{6} = \frac{5}{6}$
2 $\frac{2}{4} + \frac{3}{4} = \frac{5}{4}$
3 $\frac{5}{6} + \frac{2}{6} = \frac{7}{6}$
4 $\frac{5}{8} + \frac{2}{8} = \frac{7}{8}$
5 $\frac{1}{6} + \frac{3}{6} = \frac{4}{6}$
6 $\frac{3}{8} + \frac{2}{8} = \frac{5}{8}$
7 $\frac{6}{10} + \frac{2}{10} = \frac{8}{10}$
8 $\frac{4}{8} + \frac{3}{8} = \frac{7}{8}$
9 $\frac{2}{6} + \frac{3}{6} = \frac{5}{6}$
10 $\frac{3}{10} + \frac{5}{10} = \frac{8}{10}$
11 $\frac{2}{6} + \frac{5}{6} = \frac{7}{6}$
12 $\frac{1}{6} + \frac{10}{6} = \frac{11}{6}$

UNIT 3 STATISTICS & PROBABILITY

Spin a Number

1 Jade
2 Leo
3 Once
4 Teacher check
5 Twice
6 Teacher check

UNIT 3 MEASUREMENT & SPACE

Slides, Flips and Turn

1 
flip

2 
turn

3 
flip

4 
turn (180°)

5 
flip and turn

UNIT 3 NAPLAN*-STYLE QUESTIONS

1 $1\frac{1}{8}$

2 

3 $\frac{1}{3}$

UNIT 3 PROBLEM OF THE WEEK

Move it The solution to the problem is open-ended.

Further Investigation

The answers make a checkerboard:
For an even number of moves, the tag ends on a yellow square and for an odd number of moves, the tag lands on a white square.

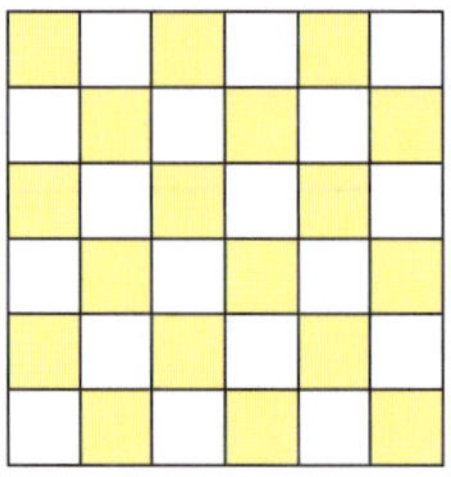

UNIT 4 NUMBER & ALGEBRA

Percentages

1 8%, $\frac{8}{100}$ and 0·08
2 80%, $\frac{8}{10}$ and 0·8
3 28%, $\frac{28}{100}$ and 0·28
4 5%, $\frac{5}{100}$ and 0·05
5 60%, $\frac{6}{10}$ and 0·6
6 36%, $\frac{36}{100}$ and 0·36

Quick Check

1 0·03 and 3%
2 0·22 and 22%
3 0·05 and 5%
4 0·17 and 17%
5 0·18 and $\frac{18}{100}$
6 0·06 and $\frac{6}{100}$
7 0·45 and $\frac{45}{100}$
8 0·09 and $\frac{9}{100}$

Percentage Discounts

1 0·2 or $\frac{20}{100}$, discount \$20, cost \$80
2 0·15 or $\frac{15}{100}$, discount \$15, cost \$85
3 0·07 or $\frac{7}{100}$, discount \$7, cost \$93
4 0·5 or $\frac{50}{100}$, discount \$50, cost \$50
5 0·3 or $\frac{30}{100}$, discount \$30, cost \$70
6 0·1 or $\frac{10}{100}$, discount \$10, cost \$90
7 0·12 or $\frac{12}{100}$, discount \$12, cost \$88
8 0·09 or $\frac{9}{100}$, discount \$9, cost \$91

UNIT 4 STATISTICS & PROBABILITY

The Fruit Survey

1 

2 Mandarin
3 25 people
4 35%
5 65%

UNIT 4 MEASUREMENT & SPACE

Area and Perimeter

1 Area 15 m^2 and Perimeter 16 m
2 Area 24 m^2 and Perimeter 20 m
3 Area 35 m^2 and Perimeter 24 m
4 Area 24 m^2 and Perimeter 22 m

UNIT 4 NAPLAN*-STYLE QUESTIONS

1 0·23
2 Area 27 m^2 and Perimeter 24 m
3 12·5% or $12\frac{1}{2}$%

UNIT 4 PROBLEM OF THE WEEK

Counters on a Grid

Miro's grid:

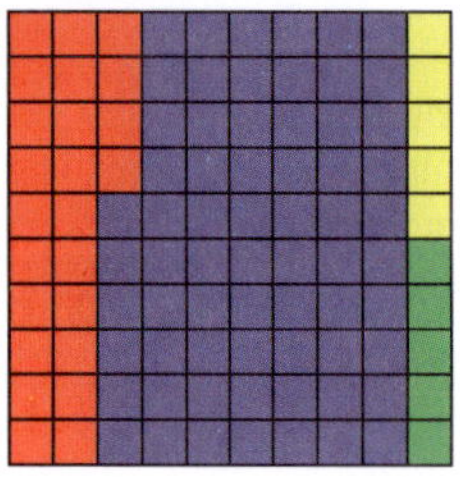

Ella's grid:

Own Counters on a Grid solution is open-ended.

UNIT 5 NUMBER & ALGEBRA

Multiplication of Large Numbers

1

	70	+ 3	
			= 73
× 8	560	24	= 584

Between 560 and 640

2

	200	+ 20	+ 2	
				= 222
× 6	1200	120	12	= 1332

Between 1320 and 1380

3

	30	+ 6	
			= 36
× 5	150	30	= 180

Between 150 and 200

ANSWERS

TARGETING MATHS HOMEWORK: YEAR 5 © PASCAL PRESS ISBN: 9781925726589

4

	100	30	5	= 135
× 8	800	240	40	= 1080

Between 1040 and 1120

5

	80	3	= 83
× 6	480	18	= 498

Between 480 and 540

6

	400	70	6	= 476
× 5	2000	350	30	= 2380

Between 2350 and 2400

Quick Check

1 Between 240 and 300
2 Between 640 and 720
3 Between 350 and 420
4 Between 150 and 200
5 Between 360 and 420
6 Between 180 and 270
7 Between 480 and 540
8 Between 280 and 320
9 Between 270 and 300
10 Between 120 and 160

Multiplying Money Amounts

1

	$5	70c	5c	= $5·75
× 5	$25	$3·50	25c	= $28·75

Between $25 and $30

2

	$2	20c	5c	= $2·25
× 7	$14	$1·40	35c	= $15·75

Between $14 and $21

3

	$6	30c	9c	= $6·39
× 4	$24	$1·20	36c	= $25·56

Between $24 and $28

4

	$8	30c	7c	= $8·37
× 6	$48	$1·80	42c	= $50·22

Between $48 and $54

5

	$12	60c	5c	= $12·65
× 3	$36	$1·80	15c	= $37·95

Between $36 and $39

6

	$3	20c	5c	= $3·25
× 9	$27	$1·80	45c	= $29·25

Between $27 and $36

UNIT 5 STATISTICS & PROBABILITY

Just by Chance

1 7
2 Heads: 6 out of 10. Tails: 4 out of 10.
3 Maybe
4 Teacher check
5 Yes, because 5 out of 10 is the same as 1 out of 2.

UNIT 5 MEASUREMENT & SPACE

Coordinate Grids

1, 3, 5

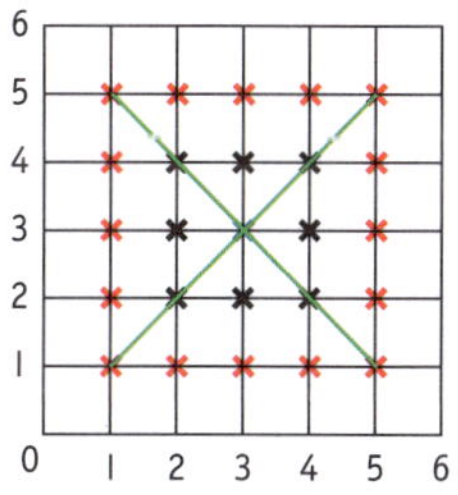

2 (2, 4) (3, 4) (4, 4) (4, 3) (4, 2) (3, 2) (2, 2) (2, 3)
4 (1, 5) (2, 5) (3, 5) (4, 5) (5, 5) (5, 4) (5, 3) (5, 2) (5, 1) (4, 1) (3, 1) (2, 1) (1, 1) (1, 2) (1, 3) (1, 4)

UNIT 5 NAPLAN*-STYLE QUESTIONS

1 $8·10
2 5 in 8
3 (1, 4)

UNIT 5 PROBLEM OF THE WEEK

Prizes for the Lucky Dip

	$7	40c	5c	= $7·45
× 3	$21	$1·20	15c	= $22·35

	$3	50c	5c	= $3·55
× 6	$18	$3·00	30c	= $21·30

	$7	40c	5c	= $7·45
× 4	$28	$1·60	20c	= $29·80

	$5	30c	9c	= $5·39
× 5	$25	$1·50	45c	= $26·95

Jade spent $21·30 on the bouncy balls.
Leo spent $29·80 on the notebooks.
Miro spent $22·35 on the pens.
Ella spent $26·95 on the pencil sharpeners.

UNIT 6 NUMBER & ALGEBRA

Division and Related Multiplication Facts

1 4, 4
2 9, 9
3 8, 8
4 7, 7
5 9, 9
6 8, 8
7 7, 7
8 6, 6
9 8, 8
10 7, 7
11 9, 9
12 9, 9
13 6, 6
14 5, 5

Quick Check

1 1
2 9
3 3
4 7
5 11
6 4
7 7
8 7
9 1
10 8

Extending Division Facts

1 450 ÷ 5
45 ÷ 5 = 9
450 ÷ 5 = 9 × 10 = 90

2 720 ÷ 8
72 ÷ 8 = 9
720 ÷ 8 = 9 × 10 = 90

3 320 ÷ 8
32 ÷ 8 = 4
320 ÷ 8 = 4 × 10 = 40

4 720 ÷ 9
72 ÷ 9 = 8
720 ÷ 9 = 8 × 10 = 80

5 560 ÷ 7
56 ÷ 7 = 8
560 ÷ 7 = 8 × 10 = 80

6 810 ÷ 9
81 ÷ 9 = 9
810 ÷ 9 = 9 × 10 = 90

7 540 ÷ 9
54 ÷ 9 = 6
540 ÷ 9 = 6 × 10 = 60

8 420 ÷ 6
42 ÷ 6 = 7
420 ÷ 6 = 7 × 10 = 70

UNIT 6 STATISTICS & PROBABILITY

Measuring Chance Events

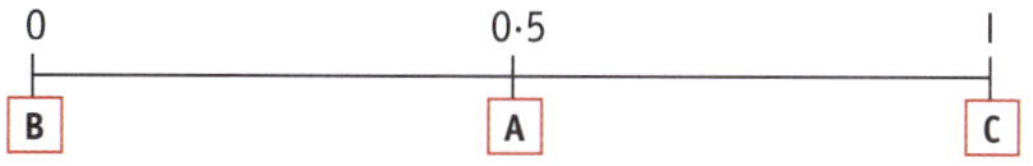

D-E Open-ended

UNIT 6 MEASUREMENT & SPACE

24-hour Time

1 15:30
2 16:20
3 05:45
4 10:15
5 09:35
6 19:20
7 20:50
8 07:45
9 15:40
10 07:35

UNIT 6 NAPLAN*-STYLE QUESTIONS

1 8
2 13:45
3 C

ANSWERS

UNIT 6 PROBLEM OF THE WEEK

Calculator Magic

Every time, the result should be the original number.
If the number was, say, 23, then 23 × 8 + 23 is the same as 23 × 9 and ÷ 9 brings us back to 23.

For the second part, using 23 as the chosen number, 23 = 20 + 3 and 32 = 30 + 2. When we add, we get 20 + 3 + 30 + 2 = 22 + 33. Divide by 11 to get 2 + 3 = 5.
The result is the sum of the two chosen digits.

UNIT 7 NUMBER & ALGEBRA

Rounding

1 Down to 150, Up to 200
2 Up to 430, Down to 400
3 Up to 350, Down to 300
4 Up to 970, Up to 1000
5 Up to 1·4, Down to 1·36
6 Up to 2·8, Up to 2·76
7 Up to 4·9, Down to 4·86
8 Down to 3·2, Up to 3·23
9 Up to $1·45, Down to $1·40
10 Up to $3·60, Up to $3·60

Quick Check

1 Up to 15
2 Down to 13
3 Up to $2
4 Down to $2
5 Down to $17
6 Down to 15
7 Down to 5
8 Up to $39
9 Up to $6
10 Up to $10

Round and Adjust

1 $78^{+2} + 63^{-2}$
= 80 + 61
= 141

2 $37^{+3} + 89^{-3}$
= 40 + 86
= 126

3 $43^{-1} + 69^{+1}$
= 42 + 70
= 112

4 $7·75 $^{+25c}$ + $3·45 $^{-25c}$
= $8 + $3·20
= $11·20

5 $16·89 $^{+11c}$ + $3·25 $^{-11c}$
= $17 + $3·14
= $20·14

6 $27·35 $^{-20c}$ + $12·80 $^{+20c}$
= $27·15 + $13
= $40·15

UNIT 7 STATISTICS & PROBABILITY

The Bean Growth Science Experiment

1 Bean A by 3 cm
2 Wednesday to Thursday
3 Thursday to Friday
4 2 cm
5 4 cm

UNIT 7 MEASUREMENT & SPACE

Draw Me a House

1–7 The diagram should look like:

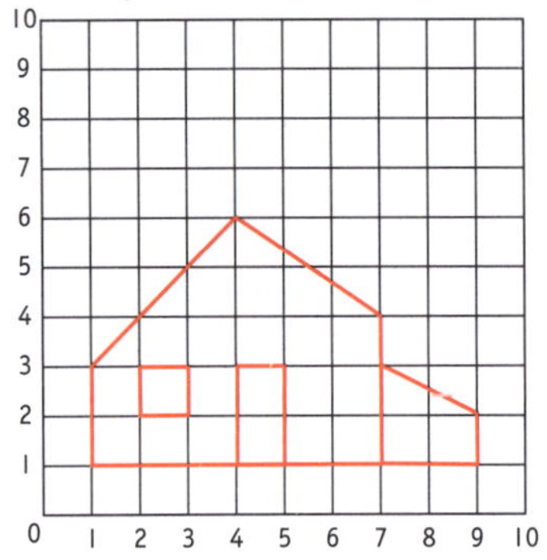

8–9 Open-ended

UNIT 7 NAPLAN*-STYLE QUESTIONS

1 $12
2 The first chart
3 (3, 2)

UNIT 7 PROBLEM OF THE WEEK

Training Data

Asha's data is shown in blue.
Ella's data is shown in orange.
Ravi's data is shown in purple.
Will's data is shown in green.
The rest is open-ended.

UNIT 8 NUMBER & ALGEBRA

Remainders

1 2 r 2
2 4 r 1
3 3 r 2
4 4 r 1
5 7 r 4
6 6 r 2
7 $3 r $1, $13 ÷ 4 = $3·25
8 $7 r $1, $15 ÷ 2 = $7·50
9 $3 r $1, $16 ÷ 5 = $3·20
10 $6 r $2, $26 ÷ 4 = $6·50

Quick Check

1 5 r 1
2 5 r 3
3 4 r 3
4 5 r 3
5 10 r 1
6 $6·50
7 $5·50
8 $4·75
9 $5·60
10 $10·10

Division Word Problems

1 $9 ÷ 2 = $4·50
2 $6 ÷ 4 = $1·50
3 $32 ÷ 5 = $6·40
4 $122 ÷ 4 = $30·50
5 $27 ÷ 5 = $5·40

UNIT 8 STATISTICS & PROBABILITY

At the Movies

1 80
2 100
3 380
4 Open-ended, but more than 120.
5 Open-ended

UNIT 8 MEASUREMENT & SPACE

Measuring and Constructing Angles

1 65° and acute
2 135° and obtuse
3 235° and reflex
4–6 Use a protractor to check.

UNIT 8 NAPLAN*-STYLE QUESTIONS

1 6
2 30
3 Second diagram

UNIT 8 PROBLEM OF THE WEEK

Magic Hat Tricks

When the tables are completed, the matching results give the answer.

Hats	1	2	3	4	5	6	7	8
Rabbits	3	5	7	9	11	13	15	17

Hats	1	2	3	4	5	6	7	8
Rabbits			3	6	9	12	15	18

The mathemagician has 7 hats and 15 rabbits.

Birds and Bird Hats

Bird-hats	1	2	3	4	5	6	7	8	9	10	11	12	13	14
Birds	5	8	11	14	17	20	23	26	29	32	35	38	41	44

Bird-hats	1	2	3	4	5	6	7	8	9	10	11	12	13	14
Birds			0	4	8	12	16	20	24	28	32	36	40	44

The mathemagician has 14 bird-hats and 44 birds.

TERM 1 REVIEW

Review: Unit 1 and Unit 2

1 $\frac{3}{10} = 0·3$
2 $\frac{14}{100} = 0·14$
3 $\frac{5}{100} = 0·05$
4 $\frac{6}{1000} = 0·006$
5 $\frac{13}{1000} = 0·013$
6 $\frac{16}{100} = 0·16$
7 $0·9 + 0·6 = \frac{9}{10} + \frac{6}{10} = \frac{15}{10}$
or 1·5
8 $0·03 + 0·08 = \frac{3}{100} + \frac{8}{100}$
$= \frac{11}{100}$ or 0·11
9 0·004 + 0·005
$= \frac{4}{1000} + \frac{5}{1000} = \frac{9}{1000}$ or 0·009
10 0·008 + 0·007
$= \frac{8}{1000} + \frac{7}{1000} = \frac{15}{1000}$ or 0·015
11 Area of a rug, B, m^2
12 Capacity of a jug, D, mL
13 Mass of a sack of potatoes, A, kg

TARGETING MATHS HOMEWORK: YEAR 5 © PASCAL PRESS ISBN: 9781925726589

14 Width of this page, C, cm
15 Mass of a blueberry, E, g
16 1 and 5
17 1, 2, 3, 4, 6 and 12
18 1 and 13
19 1, 2, 4, 8 and 16
20 Fish and Chips
21 Burger
22 D
23 B
24 11

Review: Unit 3 and Unit 4

1 $1\frac{3}{5} = \frac{8}{5}$
2 $3\frac{1}{4} = \frac{13}{4}$
3 $1\frac{3}{10} = \frac{13}{10}$
4 $2\frac{5}{8} = \frac{21}{8}$
5 $2\frac{1}{2} = \frac{5}{2}$
6 $4\frac{2}{3} = \frac{14}{3}$
7

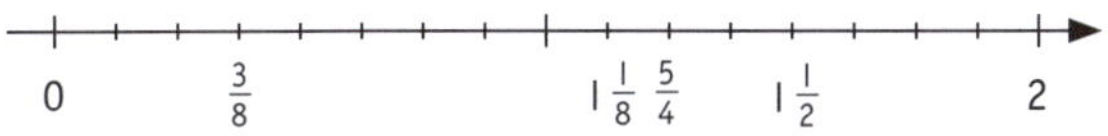

8 $1\frac{3}{4}$
9 $1\frac{1}{5}$
10 $\frac{1}{6} + \frac{2}{6} + \frac{3}{6} = 1$
11 $\frac{2}{8} + \frac{3}{8} + \frac{6}{8} = \frac{11}{8} = 1\frac{3}{8}$
12 Second spinner
13 0·06 and 6%
14 0·24 and 24%
15 0·7 and 70%
16 $80
17 $85
18 $92
19 Area = 28 m^2
Perimeter = 22 m
20 Area = 36 m^2
Perimeter = 26 m

Review: Unit 5 and Unit 6

1

	50	+ 6	= 56
× 5	250	30	= 280

2

	80	+ 3	= 83
× 7	560	21	= 581

3 Between 240 and 300
4 Between 490 and 560
5

	$7	+ 80c	+ 5c	= $7·85
× 6	$42	$4·80	30c	= $47·10

6

	$6	+ 30c	+ 8c	= $6·38
× 4	$24	$1·20	32c	= $25·52

7 3 heads out of 8 tosses
8 5 tails out of 8 tosses
9 (1, 2) (2, 3), (3, 3), (4, 2), (3, 1) (2, 1)
10 45 ÷ 9 = 5
5 × 9 = 45
11 72 ÷ 8 = 9
9 × 8 = 72
12 320 ÷ 8
32 ÷ 8 = 4
320 ÷ 8 = 10 × 4 = 40
13 540 ÷ 9
54 ÷ 9 = 6
540 ÷ 9 = 10 × 6 = 60
14 10:50
15 18:45
16 23:55

Review: Unit 7 and Unit 8

1 3·8
2 2·8
3 4
4 1·77
5 1·35
6 1·27
7 $1·70
8 $3·70
9 $3
10 $3
11 $77^{+3} + 26^{-3}$
= 80 + 23
= 103
12 $56^{+4} + 75^{-4}$
= 60 + 71
= 131
13 $42^{-2} + 58^{+2}$
= 40 + 60
= 100
14 50 mg
15 250 mg
16 14 ÷ 4 = 3 r 2
17 27 ÷ 5 = 5 r 2
18 $21 ÷ 4 = $5·25
19 $17 ÷ 5 = $3·40
20 90 people
21 15 people
22 90° right
23 45° acute
24 180° straight

TERM 2

UNIT 9 NUMBER & ALGEBRA

Finding Multiples

1 60
2 12
3 60
4 30
5 300
6 72
7 84
8 80
9 48
10 72
11 30
12 36
13 42
14 12
15 24
16 24
17 20
18 18
19 42
20 72

Quick Check

1 20
2 78
3 6
4 78
5 12
6 6
7 24
8 30
9 14
10 60

Decimals on a Number Line

1 0·6
2 0·15
3 0·34
4 0·015
5 0·028
6 0·0012

UNIT 9 STATISTICS & PROBABILITY

Dice Totals

1

		Dice 1					
		1	2	3	4	5	6
Dice 2	1	2	3	4	5	6	7
	2	3	4	5	6	7	8
	3	4	5	6	7	8	9
	4	5	6	7	8	9	10
	5	6	7	8	9	10	11
	6	7	8	9	10	11	12

2 2 and 12
3 7
4 Open-ended (1 point for 6, 7 or 8)
5 Open-ended (1 point for 'Because that is one of the most likely to occur.')

UNIT 9 MEASUREMENT & SPACE

Scale it Up

1–6 1 point for each item correctly scaled up.

UNIT 9 NAPLAN*-STYLE QUESTIONS

1 0·75
2 250 mm
3 $\frac{1}{3}$

UNIT 9 PROBLEM OF THE WEEK

Two-dice Bingo

1–2 Ella and Will have the best chance of winning because their chosen numbers have a much more likely chance of occurring than a 2.
3 The results are:

Ella

~~7~~	~~5~~	~~7~~
~~6~~	7	~~4~~

Will

~~5~~	6	~~7~~
~~8~~	~~7~~	~~6~~

Asha

~~2~~	~~4~~	~~6~~
~~3~~	~~5~~	~~7~~

Ravi

~~2~~	~~7~~	~~6~~
~~9~~	6	~~5~~

5 Asha had two totals that were unlikely to occur.
6–7 Open-ended

UNIT 10 NUMBER & ALGEBRA

Truth Sentences

1 ≠
2 =
3 =
4 ≠
5 =
6 =
7 ≠
8 ≠
9 =
10 =
11 ≠
12 ≠
13 ≠
14 ≠
15 ≠
16 =
17 =
18 =
19 ≠
20 =

Quick Check

1 3 × 4
2 2 × 8
3 4 × 7
4 9 × 4
5 7 × 8
6 9 × 9
7 9 × 7
8 8 × 7
9 11 × 3
10 9 × 12

Using Factors for Mental Multiplication

1 23 × 2 = 46, 46 × 2 = 92
2 23 × 2 = 46, 46 × 3 = 138
3 23 × 4 = 92, 92 × 2 = 184
4 Use 9 = 3 × 3, 23 × 3 = 69, 69 × 3 = 207
5 Use 4 = 2 × 2, 32 × 2 = 64, 64 × 2 = 128
6 Use 6 = 2 × 3, 32 × 2 = 64, 64 × 3 = 192
7 Use 9 = 3 × 3, 32 × 3 = 96, 96 × 3 = 288

UNIT 10 STATISTICS & PROBABILITY

The Bean Experiment

1 Graph A
2 Same
3 14 cm
4 10 cm
5 Graph A

UNIT 10 MEASUREMENT & SPACE

Drawing Angles

Teacher check

ANSWERS

UNIT 10 NAPLAN*-STYLE QUESTIONS

1 48 × 2 × 2 × 2 2 The second option 3 9

UNIT 10 PROBLEM OF THE WEEK

A Symmetrical Design

Leo's design with angles marked:	With the 45° and 90° angles marked separately, it is easier to find the total of the angles:
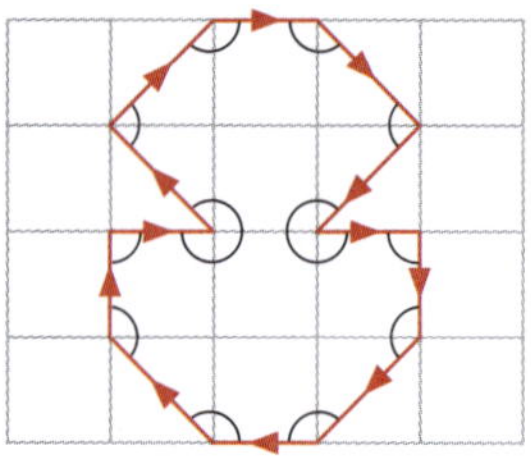	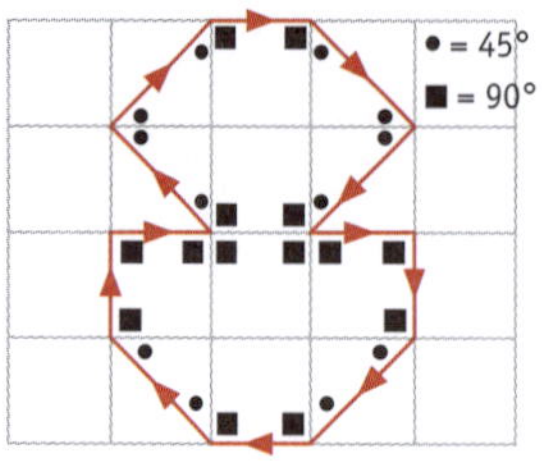
135° + 90° + 315° + 90° + 135° + 135° + 135° + 90° + 315° + 90° + 135° = 1800°	14 × 90° + 12 × 45° = 1260° + 540° = 1800°

UNIT 11 NUMBER & ALGEBRA

Fractions

1 $\frac{1}{4}$	4 $\frac{1}{3}$	7 $\frac{1}{5}$	10 $\frac{1}{5}$	13 $\frac{1}{4}$
2 $\frac{1}{8}$	5 $\frac{1}{6}$	8 $\frac{1}{10}$	11 $\frac{2}{3}$	14 $\frac{3}{4}$
3 $\frac{1}{16}$	6 $\frac{1}{12}$	9 $\frac{1}{20}$	12 $\frac{1}{5}$	15 $\frac{1}{2}$

Quick Check

1 $\frac{1}{4}$	3 $\frac{5}{8}$	5 $\frac{7}{10}$	7 $\frac{2}{3}$	9 $\frac{1}{2}$
2 $\frac{3}{5}$	4 $\frac{7}{12}$	6 $\frac{1}{2}$	8 $\frac{1}{3}$	10 $\frac{1}{1}$ or 1

Percentages

1 40%	3 89%	5 20%
2 25%	4 75%	6 76%

UNIT 11 STATISTICS & PROBABILITY

Graphing Statistics

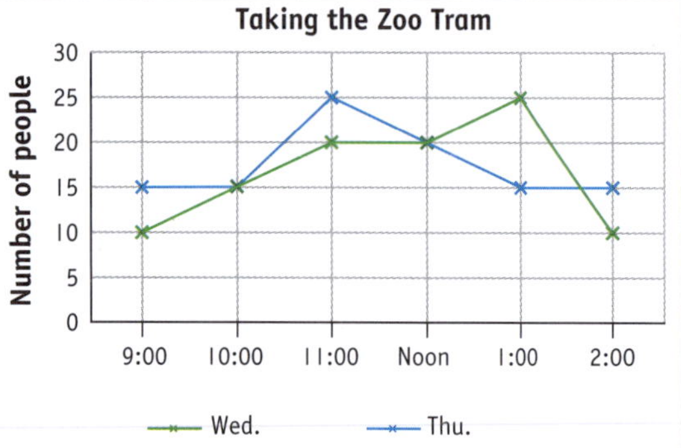

UNIT 11 MEASUREMENT & SPACE

Nets in the Real World

1 cube, Rubik's cube
2 triangular prism, Toblerone box
3 cone, ice-cream cone
4 cylinder, tin
5 rectangular prism, refrigerator

UNIT 11 NAPLAN*-STYLE QUESTIONS

1 2 triangles and 3 rectangles
2 45%
3 6 choc drops

UNIT 11 PROBLEM OF THE WEEK

Fill the Board

1 Ella
2 Miro: 25%
Ella: 10% and 5%
Asha: 10% and 10%
Sam: 10% and 10%
3–4 Open-ended

UNIT 12 NUMBER & ALGEBRA

Addition of Fractions

1–2 $\frac{3}{6} = \frac{1}{2}$ 7–8 $\frac{8}{10} = \frac{4}{5}$
3–4 $\frac{5}{10} = \frac{1}{2}$ 9–10 $\frac{9}{8} = 1\frac{1}{8}$
5–6 $\frac{5}{8}$

Quick Check

1 $\frac{2}{4}$ or $\frac{1}{2}$	3 $\frac{4}{4}$ or 1	5 $\frac{5}{10}$ or $\frac{1}{2}$	7 $\frac{7}{6}$ or $1\frac{1}{6}$	9 $\frac{6}{10}$ or $\frac{3}{5}$
2 $\frac{5}{5}$ or 1	4 $\frac{6}{8}$ or $\frac{3}{4}$	6 $\frac{10}{10}$ or 1	8 $\frac{9}{8}$ or $1\frac{1}{8}$	10 $\frac{6}{6}$ or 1

Subtraction of Fractions

1 $\frac{4}{6} - \frac{1}{6} = \frac{3}{6}$ or $\frac{1}{2}$
2 $\frac{3}{8} - \frac{2}{8} = \frac{1}{8}$
3 $\frac{5}{10} - \frac{3}{10} = \frac{2}{10}$ or $\frac{1}{5}$
4 $\frac{6}{8} - \frac{3}{8} = \frac{3}{8}$
5 $\frac{4}{10} - \frac{1}{10} = \frac{3}{10}$
6 $\frac{5}{8} - \frac{2}{8} = \frac{3}{8}$
7 $\frac{6}{10} - \frac{3}{10} = \frac{3}{10}$
8 $\frac{6}{8} - \frac{1}{8} = \frac{5}{8}$
9 $\frac{6}{6} - \frac{5}{6} = \frac{1}{6}$
10 $\frac{9}{12} - \frac{5}{12} = \frac{4}{12}$ or $\frac{1}{3}$

UNIT 12 STATISTICS & PROBABILITY

Fractions for Chance

1 4 in 5
2 1 in 5
3 0 in 5 – impossible
4 2 in 5
5 5 in 5 – certain

UNIT 12 MEASUREMENT & SPACE

Area and Perimeter

1 8 + 3 + 4 + 3 + 12 + 6 = 36 cm
2 48 + 12 = 60 cm²
3 Yes
4 No
5 72 cm
6 240 cm²

UNIT 12 NAPLAN*-STYLE QUESTIONS

1 12 2 First shape 3 Third spinner

UNIT 12 PROBLEM OF THE WEEK

The Pizza Party

$\frac{2}{8}$ or $\frac{1}{4}$ of a pizza

UNIT 13 NUMBER & ALGEBRA

Division with Remainders

1 5 r 1 2 3 r 3 3 6 r 2 4 6 r 1
5 15 ÷ 4 = 3 r 3 so 3 oranges left over.
6 37 ÷ 5 = 7 r 2 so 7 nets will be needed.
7 26 ÷ 5 = 5 r 1 so 1 orange was left over.
8 27 ÷ 6 = 4 r 3 so 4 nets can be made.
9 50 ÷ 8 = 6 r 2 so there were 50 avocadoes altogether.

Quick Check

1 7 r 2	3 9 r 1	5 7 r 1	7 12 r 1	9 7 r 2
2 4 r 6	4 9 r 2	6 8 r 1	8 10 r 3	10 9 r 5

Division with Fraction Remainders

1 $5 \div 2 = 2\frac{1}{2}$ 3 $4 \div 3 = 1\frac{1}{3}$ 5 $7 \div 4 = 1\frac{3}{4}$
2 $5 \div 4 = 1\frac{1}{4}$ 4 $9 \div 6 = 1\frac{1}{2}$

ANSWERS

TARGETING MATHS HOMEWORK: YEAR 5 © PASCAL PRESS ISBN: 9781925726589

UNIT 13 STATISTICS & PROBABILITY

Sport on TV

1

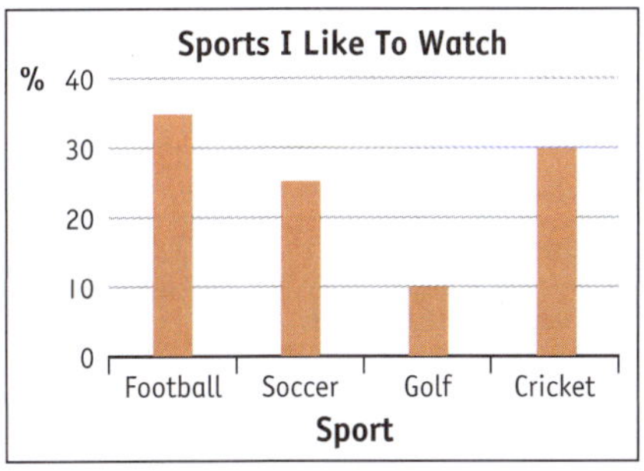

2 35 3 25 4 10 5 30

UNIT 13 MEASUREMENT & SPACE

Capacity

1 90 mL
2 120 mL
3 $\frac{1}{4}$ cup and $\frac{1}{3}$ cup
4 $\frac{1}{4}$ cup and $\frac{1}{2}$ cup
5 $\frac{1}{2}$ cup and two $\frac{1}{3}$ cups

UNIT 13 NAPLAN*-STYLE QUESTIONS

1 1 strawberry 2 60 mL 3 20%

UNIT 13 PROBLEM OF THE WEEK

Calculator Magic

An example	
2	7
Memory	27
	9
8	11
7	13
6	15
5	17
4	19
3	21
2	23
1	25
Display	27

The display number is always the same as the starting number.

UNIT 14 NUMBER & ALGEBRA

Round and Adjust

1 $8·75 × 6
↓ Rounding by adding 25c
$9 × 6 = $54
↓ Adjust by subtracting 25c × 6 = $1·50
$52·50

2 $10·80 × 5
↓ Rounding by adding 20c
$11 × 5 = $55
↓ Adjust by subtracting 20c × 5 = $1
$54

3 $3·90 × 7
↓ Rounding by adding 10c
$4 × 7 = $28
↓ Adjust by subtracting 10c × 7 = 70c
$27·30

4 $11·80 × 8
↓ Rounding by adding 20c
$12 × 8 = $96
↓ Adjust by subtracting 20c × 8 = $1·60
$94·40

5 $9·05 × 7
↓ Rounding by subtracting 5c
$9 × 7 = $63
↓ Adjust by adding 5c × 7 = 35c
$63·35

6 $12·30 × 4
↓ Rounding by subtracting 30c
$12 × 4 = $48
↓ Adjust by adding 30c × 4 = $1·20
$49·20

Quick Check

1 $29·40 3 $20·65 5 $38·40 7 $13 9 $20·75
2 $11·40 4 $28·50 6 $18·45 8 $7·20 10 $28

Practical Word Problems

1 95c × 4 = $3·80, $5 − $3·80 = $1·20
2 $15 × 4 + $20 = $80, $100 − $80 = $20
3 $24 ÷ 4 = $6, $6 × 3 = $18
4 24 × $\frac{1}{4}$ = 6, 6 × $\frac{1}{2}$ = 3
5 $3·50 × 5 = $17·50, $17·50 + $12·50 = $30

UNIT 14 STATISTICS & PROBABILITY

Finding the mode

1 The mode is 4.
2 The mode is 5.
3 The mode is 2.
4 The mode is 3.
5 The mode is 3.

UNIT 14 MEASUREMENT & SPACE

Calling Times

1 15:00
2 05:10
3 21:20
4 13:00
5 14:00
6 03:00
7 08:00
8 Adelaide time 7:50 am or 8:00 am

UNIT 14 NAPLAN*-STYLE QUESTIONS

1 $1·75 2 3 h 38 min 3 75 cm

UNIT 14 PROBLEM OF THE WEEK

Reading the Paper

Day	Time (minutes)
Monday	10
Tuesday	5
Wednesday	30
Thursday	15
Friday	25
Saturday	5
Sunday	30
Total	120

UNIT 15 NUMBER & ALGEBRA

Multiplication and Division with Unknowns

1 4 × 3 = 36 ÷ 3 = 12
2 4 × 7 = 56 ÷ 2 = 28
3 54 ÷ 3 = 3 × 6 = 18
4 4 × 8 = 64 ÷ 2 = 32
5 110 ÷ 10 = 11 × 1 = 11
6 72 ÷ 12 = 48 ÷ 8 = 6
7 13 × 5 = 130 ÷ 2 = 65
8 3 × 20 = 180 ÷ 3 = 60
9 9 × 6 = 108 ÷ 2 = 54
10 18 × 2 = 108 ÷ 3 = 36
11 99 ÷ 9 = 88 ÷ 8 = 11
12 88 ÷ 4 = 66 ÷ 3 = 22

Quick Check

Sample answers – other answers are possible.

1 3 × 2 = 12 ÷ 2
2 4 × 3 = 24 ÷ 2
3 3 × 6 = 36 ÷ 2
4 3 × 5 = 45 ÷ 3
5 4 × 4 = 32 ÷ 2
6 9 × 3 = 54 ÷ 2
7 7 × 2 = 56 ÷ 4
8 4 × 7 = 56 ÷ 2
9 7 × 4 = 84 ÷ 3
10 2 × 2 = 36 ÷ 9

ANSWERS

Factors

	21	22	23	24	25	26	27	28	29
2		✓		✓		✓		✓	
3	✓			✓			✓		
4				✓				✓	
5					✓				
6				✓					
7	✓							✓	
8				✓					
9							✓		
10									
11		✓							
12				✓					
13						✓			
14								✓	
Total	2	2	0	6	1	2	2	4	0

2 24
3 21, 22, 26, 27
4 21, 28
5 25
6 2, 3, 4, 6, 8, 12
7 23, 29

UNIT 15 MEASUREMENT & SPACE

Coordinates

1. To collect the at (1, 1) from (0, 0), go right 1 and up 1.
2. To collect the at (1, 5) from the at (1, 1), go up 4.
3. To collect the at (2, 4) from the at (1, 5), go right 1, down 1.
4. To collect the at (5, 3) from the at (2, 4), go right 3, down 1.
5. To reach the exit at (6, 6) from the at (5, 3), go right 1, up 3.

UNIT 15 NAPLAN*-STYLE QUESTIONS

1 10 **2** 15 **3** (2, 1) and (4, 2)

UNIT 15 PROBLEM OF THE WEEK

Collect the Treasure and Escape

1

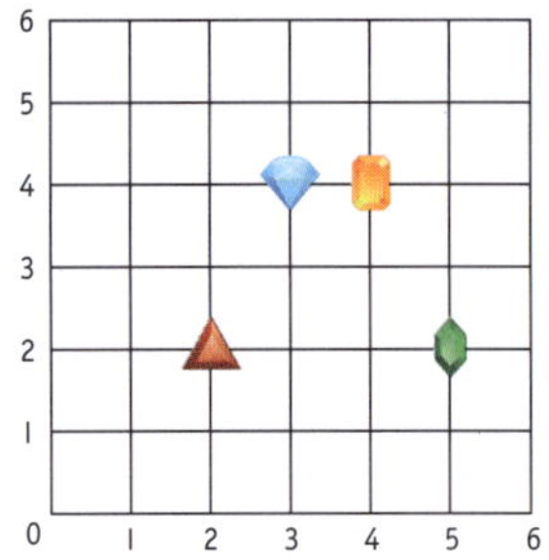

The remaining questions are open-ended.

UNIT 16 NUMBER & ALGEBRA

Decimals as Fractions

1. $\frac{27}{100}$ or $\frac{2}{10} + \frac{7}{100}$
2. $\frac{76}{100}$ or $\frac{7}{10} + \frac{6}{100}$
3. $\frac{138}{1000}$ or $\frac{1}{10} + \frac{3}{100} + \frac{8}{1000}$
4. $\frac{348}{1000}$ or $\frac{3}{10} + \frac{4}{100} + \frac{8}{1000}$
5. $\frac{257}{1000}$ or $\frac{2}{10} + \frac{5}{100} + \frac{7}{1000}$
6. $\frac{158}{1000}$ or $\frac{1}{10} + \frac{5}{100} + \frac{8}{1000}$
7. $\frac{708}{1000}$ or $\frac{7}{10} + \frac{0}{100} + \frac{8}{1000}$
8. $\frac{109}{1000}$ or $\frac{1}{10} + \frac{0}{100} + \frac{9}{1000}$
9. $\frac{18}{1000}$ or $\frac{0}{10} + \frac{1}{100} + \frac{8}{1000}$
10. $\frac{507}{1000}$ or $\frac{5}{10} + \frac{0}{100} + \frac{7}{1000}$
11. $\frac{35}{1000}$ or $\frac{0}{10} + \frac{3}{100} + \frac{5}{1000}$
12. $\frac{123}{1000}$ or $\frac{1}{10} + \frac{2}{100} + \frac{3}{1000}$
13. 0·12
14. 0·304
15. 0·056
16. 1·25
17. 3·52
18. 2·727

Quick Check

1 0·435 **2** 0·705 **3** 0·086 **4** 0·706 **5** 0·033 **6** 0·34 **7** 0·665 **8** 0·808 **9** 0·263 **10** 0·043

Equivalent Fractions

1 $\frac{8}{16}$ **2** $\frac{2}{10}$ **3** $\frac{3}{15}$ **4** $\frac{4}{12}$ **5** $\frac{3}{12}$ **6** $\frac{4}{10}$ **7** $\frac{4}{6}$ **8** $\frac{9}{18}$ **9** $\frac{9}{15}$ **10** $\frac{16}{20}$ **11** $\frac{24}{42}$ **12** $\frac{18}{48}$

UNIT 16 STATISTICS & PROBABILITY

The Vet's Busy Week

	Mon.	Tue.	Wed.	Thu.	Fri.	Totals
Cats	3	11	13	11	7	45
Dogs	3	11	12	6	10	42
Kittens	7	7	12	12	10	48
Puppies	5	6	12	12	8	43
Rabbits	4	5	12	12	7	40
Totals	22	40	61	53	42	218

Sample answers:

1. Which type of animal did the vet see most?
2. Which type of animal did the vet see least?
3. Which was the busiest day?
4. Which was the day the vet saw the least number of animals?
5. How many animals altogether visited the clinic that week?

UNIT 16 MEASUREMENT & SPACE

Finding Areas

1

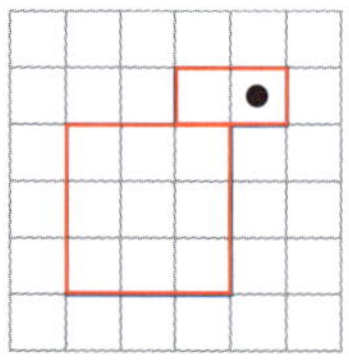

Area = 2 × 1 + 3 × 3
= 11 squares

3

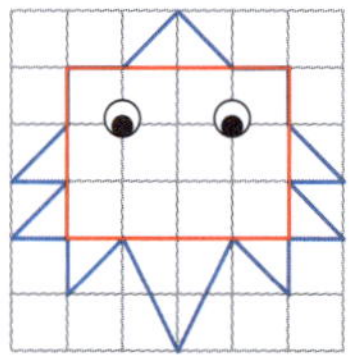

Area = 3 × 4 + 8 × $\frac{1}{2}$ + 2
= 18 squares

2

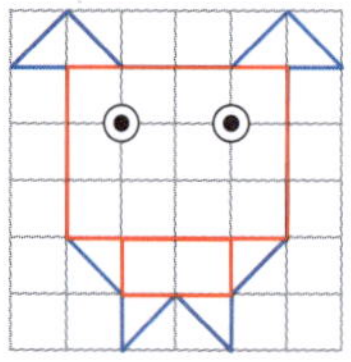

Area = 2 × 1 + 3 × 4 +
8 × $\frac{1}{2}$ = 18 squares

4

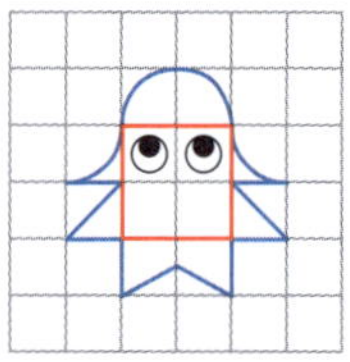

Area = 2 × 2 + 1 + 1 + 2 × $\frac{1}{2}$
+ 2 × $\frac{3}{4}$ = $8\frac{1}{2}$ squares

UNIT 16 NAPLAN*-STYLE QUESTIONS

1. 0·025
2. Second rectangle
3. Second diagram (9 circles)

UNIT 16 PROBLEM OF THE WEEK

Fundraising Cakes

Mrs Green's class have $\frac{1}{12}$ of a cake left over.

TERM 2 REVIEW

Review: Unit 9 and Unit 10

	Factors	Multiple	LCM
1	2, 4, 8	64	8
2	3, 5, 6	90	30
3	6, 5, 4	120	60

4 0·25 **5** 0·14 **6** 0·016 **7** = **8** ≠ **9** = **10** ≠ **11** = **12** ≠

The estimates of the angles are open-ended, but here are the actual answers.

13 45° **14** 135° **15** 225°

ANSWERS

16 4 = 2 × 2, 42 × 2 = 84, 84 × 2 = 168
17 6 = 2 × 3, 42 × 2 = 84, 84 × 3 = 252
18 9 = 3 × 3, 42 × 3 = 126, 126 × 3 = 378

Review: Unit 11 and Unit 12

1 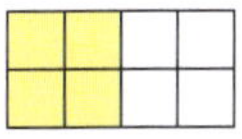
$\frac{1}{2}$

2 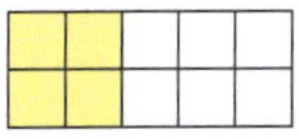
$\frac{2}{5}$

3 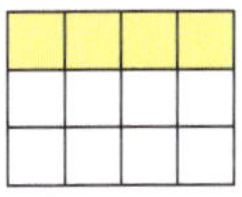
$\frac{1}{3}$

4 25%
5 79%
6 1%
7 Tuesday
8 20
9 5
10 $\frac{7}{8}$
11 $\frac{3}{6}$ or $\frac{1}{2}$
12 $\frac{1}{8}$
13 $\frac{4}{10}$ or $\frac{2}{5}$
14 $\frac{1}{6}$
15 0
16 $\frac{1}{5}$
17 $\frac{3}{5}$
18 $\frac{2}{5}$
19 $\frac{4}{5}$

Review: Unit 13 and Unit 14

1 5 r 1
2 9 r 3
3 4 r 2
4 7 r 2
5 $1\frac{3}{4}$ m
6 $2\frac{1}{6}$ kg

7–9
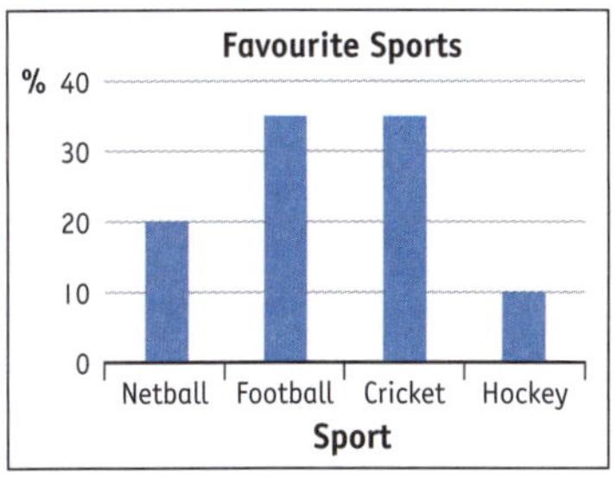

10 $15·80
11 $10·80
12 $8·05
13 $45·75
14 4
15 6
16 4:30 am in London
17 10:30 pm in London
18 1:20 pm in London

Review: Unit 15 and Unit 16

1 3 × 9 = 54 ÷ 2 = 27
2 6 × 7 = 84 ÷ 2 = 42
3 3 × 6 = 72 ÷ 4 = 18
4 12 × 4 = 96 ÷ 2 = 48
5 2, 4, 8, 16
6 2, 4, 11, 22
7 is at (1, 5)
8 is at (4, 4)
9 is at (3, 2)
10 is at (3, 6)
11 is at (5, 3)
12 is at (1, 1)
13 0·34
14 0·434
15 0·043
16 0·505
17 $\frac{3}{6}$
18 $\frac{8}{10}$
19 $\frac{1}{4}$
20 $\frac{2}{3}$
21 Totals: Cats 41, Dogs 53, Kittens 43, Puppies 40, Rabbits 48. Mon. 50, Tue. 40, Wed. 62, Thu. 41, Fri. 32. Total 225.
22 40
23 Tuesday
24 Friday
25 Open-ended, but along the lines of 'least number of customers'.

TERM 3

UNIT 17 NUMBER & ALGEBRA

Ordering Decimal Fractions

1 Decimals: 0·99, 0·56, 0·9, 1·3
Fractions: $\frac{99}{100}$, $\frac{56}{100}$, $\frac{9}{10}$, $\frac{13}{10}$
Smallest to largest: 0·56, 0·9, 0·99, 1·3

2 Decimals: 0·36, 0·99, 2·6, 1·8
Fractions: $\frac{36}{100}$, $\frac{99}{100}$, $\frac{26}{10}$, $\frac{18}{10}$
Order from smallest to largest: 0·36, 0·99, 1·8, 2·6

3 Decimals: 0·3, 0·13, 1·3, 1·35
Fractions: $\frac{3}{10}$, $\frac{13}{100}$, $\frac{13}{10}$, $\frac{135}{100}$
Order from smallest to largest: 0·13, 0·3, 1·3, 1·35

4 Decimals: 0·8, 0·18, 0·27, 0·76
Fractions: $\frac{8}{10}$, $\frac{18}{100}$, $\frac{27}{100}$, $\frac{76}{100}$
Order from smallest to largest: 0·18, 0·27, 0·76, 0·8

5 Decimals: 0·009, 0·9, 0·004, 0·45
Fractions: $\frac{9}{1000}$, $\frac{9}{10}$, $\frac{4}{1000}$, $\frac{45}{100}$
Order from smallest to largest: 0·004, 0·009, 0·45, 0·9

Quick Check

1 0·27
2 3·4
3 3·05
4 0·046
5 0·5
6 $\frac{63}{100}$
7 $\frac{36}{1000}$
8 $\frac{135}{100}$
9 $\frac{3006}{1000}$
10 $\frac{65}{1000}$

Rules for Divisibility

	The number	is divisible by 2	3	4	5	6	8	9
1	69 870	✓	✓	✗	✓	✓	✗	✗
2	34 264	✓	✗	✓	✗	✗	✓	✗
3	72 369	✗	✓	✗	✗	✗	✗	✓
4	12 645	✗	✓	✗	✓	✗	✗	✓
5	35 786	✓	✗	✗	✗	✗	✗	✗
6	45 607	✗	✗	✗	✗	✗	✗	✗
7	93 645	✗	✓	✗	✓	✗	✗	✓
8	84 279	✗	✓	✗	✗	✗	✗	✗
9	65 348	✓	✗	✓	✗	✗	✗	✗
10	8 658 345	✗	✓	✗	✓	✗	✗	✗

UNIT 17 STATISTICS & PROBABILITY

10 Counters in the Bag

1 Certain
2 Very likely
3 Likely
4 Likely
5 Unlikely

UNIT 17 MEASUREMENT & SPACE

Folding Nets

Award 5 points for choosing the one net that makes a cube with a continuous snake – Number 5 of course. And a bonus point for a correct answer to Question 6. There is a hint that students might need for this question. This method of labelling the outside edges shows which edges will join when the net is folded into a cube. It makes it easy to see how to draw a snake that is continuous.

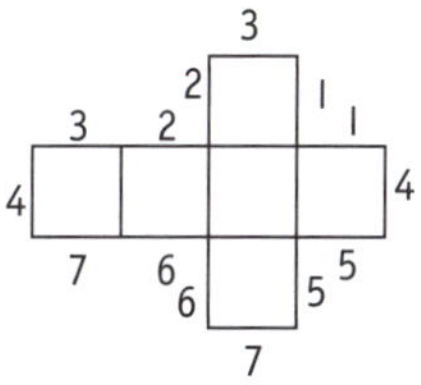

UNIT 17 NAPLAN*-STYLE QUESTIONS

1 2·333
2 9126
3 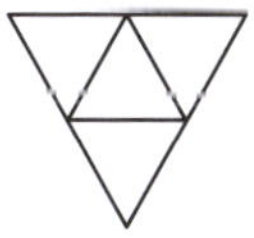

UNIT 17 PROBLEM OF THE WEEK

Making Multiples

- Asha can make 17 with 33 795 or any combination that ends with a 5.
- Ravi can make 14 at most. He has to make sure that the last two numbers of his combination make a multiple of 8, and there is an even number in the 100s place.
- Sam can make 20 with 11 358 or any combination that ends with 8.
- Leo is the winner! He can make 32 with combinations such as 79 632 and 72 936 which are both multiples of 8 and 9.

UNIT 18 NUMBER & ALGEBRA

Percentages in Daily Events

1 50%
2 25%
3 60%
4 90%
5 20%
6 80%
7 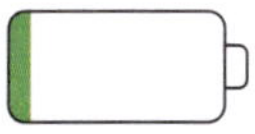
8 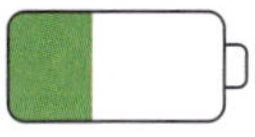
9 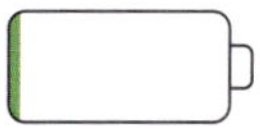
10 85%
11 75%
12 5%

ANSWERS

Quick Check

1 77%　2 6%　3 13%　4 22%　5 74%　6 84%　7 50%　8 25%　9 32%　10 68%

Fractions of a Shape

1 $\frac{2}{3}$　2 $1\frac{1}{3}$　3 $1\frac{1}{2}$　4 3　5 $\frac{2}{5}$　6 $\frac{4}{5}$

UNIT 18 STATISTICS & PROBABILITY

Broken Bones

1 56
2 Day 2 and Day 8
3 4
4 Both were the same: 28 broken bones.
5 14

UNIT 18 MEASUREMENT & SPACE

Units of Measurement

1 milligram, gram, kilogram
2 millilitre, litre
3 millimetre, centimetre, metre, kilometre

The next questions are open-ended, but here are some sample answers:

4 a road　7 chilli powder　10 sugar
5 a piece of ribbon　8 milk　11 tea
6 a medicine dose　9 a piece of wood　13 an orange

UNIT 18 NAPLAN*-STYLE QUESTIONS

1 20%　2 $\frac{2}{3}$　3 5

UNIT 18 PROBLEM OF THE WEEK

The Jeweller's Scales

Mass	Side 1	Side 2 + the metal
4 g	1 g + 3 g	
7 g	1 g + 6 g	
8 g	3 g + 6 g	1 g
9 g	3 g + 6 g	
2 g	3 g	1 g

The Challenge

Mass (g)	Weights (g) 1	3	6	13
1	✓			
2	✗	✓		
3		✓		
4	✓	✓		
5	✗		✓	
6			✓	
7	✓		✓	
8	✗	✓	✓	
9		✓	✓	
10	✓	✓	✓	
11	✓	✓	✗	✓

Mass (g)	Weights (g) 1	3	6	13
12	✗			✓
13				✓
14	✓			✓
15	✗	✓		✓
16		✓		✓
17	✓	✓		✓
18	✗		✓	✓
19			✓	✓
20	✓		✓	✓
21	✗	✓	✓	✓
22		✓	✓	✓
23	✓	✓	✓	✓

UNIT 19 NUMBER & ALGEBRA

Adding Fractions

1 $\frac{3}{8} + \frac{2}{8} = \frac{5}{8}$
2 $\frac{6}{8} + \frac{5}{8} = \frac{11}{8} = 1\frac{3}{8}$
3 $\frac{4}{6} + \frac{1}{6} = \frac{5}{6}$
4 $\frac{1}{5} + \frac{3}{10} = \frac{2}{10} + \frac{3}{10} = \frac{5}{10} = \frac{1}{2}$
5 $\frac{2}{16} + \frac{1}{16} = \frac{3}{16}$
6 $\frac{4}{8} + \frac{1}{8} = \frac{5}{8}$
7 $\frac{1}{12} + \frac{3}{12} = \frac{4}{12} = \frac{1}{3}$
8 $\frac{10}{12} + \frac{1}{12} = \frac{11}{12}$
9 $\frac{3}{8} + \frac{6}{8} = \frac{9}{8} = 1\frac{1}{8}$
10 $\frac{6}{10} + \frac{7}{10} = \frac{13}{10} = 1\frac{3}{10}$
11 $\frac{4}{5} + \frac{7}{10} = \frac{15}{10} = 1\frac{1}{2}$
12 $\frac{5}{6} + \frac{6}{12} = \frac{16}{12} = 1\frac{1}{3}$
13 $\frac{7}{8} + \frac{5}{8} = \frac{12}{8} = 1\frac{1}{2}$
14 $\frac{27}{10} + \frac{7}{5} = \frac{41}{10} = 4\frac{1}{10}$

Quick Check

1 $\frac{3}{4}$　2 $\frac{1}{2}$　3 $\frac{2}{5}$　4 2　5 $1\frac{1}{2}$　6 $\frac{1}{2}$　7 $1\frac{1}{3}$　8 $1\frac{1}{3}$　9 $1\frac{1}{10}$　10 $\frac{1}{2}$

Subtracting Fractions

1 $\frac{4}{8} - \frac{1}{8} = \frac{3}{8}$
2 $\frac{5}{8} - \frac{2}{8} = \frac{3}{8}$
3 $\frac{2}{6} - \frac{1}{6} = \frac{1}{6}$
4 $\frac{6}{8} - \frac{3}{8} = \frac{3}{8}$
5 $\frac{6}{8} - \frac{1}{8} = \frac{5}{8}$
6 $\frac{3}{6} - \frac{2}{6} = \frac{1}{6}$
7 $\frac{6}{12} - \frac{1}{12} = \frac{5}{12}$
8 $\frac{6}{16} - \frac{1}{16} = \frac{5}{16}$
9 $\frac{8}{10} - \frac{1}{10} = \frac{7}{10}$
10 $\frac{7}{8} - \frac{6}{8} = \frac{1}{8}$

UNIT 19 STATISTICS & PROBABILITY

Fitness Program Table

Sample answer:

Activities	Min	Day 1	Day 2	Day 3	Day 4	Day 5
Walking machine	20			✓	✓	✓
Weights	10		✓		✓	✓
Pilates	30	✓	✓			✓
Swimming	25	✓		✓	✓	
Rowing machine	15		✓	✓		
Stretches	5	✓	✓		✓	
Fitness Program Total		60 min	60 min	60 min	60 min	60 min

UNIT 19 MEASUREMENT & SPACE

The Landscape Garden Plan

1 26 m　2 9 m^2　3 6 m　4 29 m^2　5 26 bushes　6 6 fruit trees

UNIT 19 NAPLAN*-STYLE QUESTIONS

1 0·7　2 $\frac{5}{8}$　3 44 metres

UNIT 19 PROBLEM OF THE WEEK

Baby Animal Petting Corner

A possible layout is as shown, but other layouts are possible. However, every animal pen must be accessible from the path from the Entrance to the Exit.

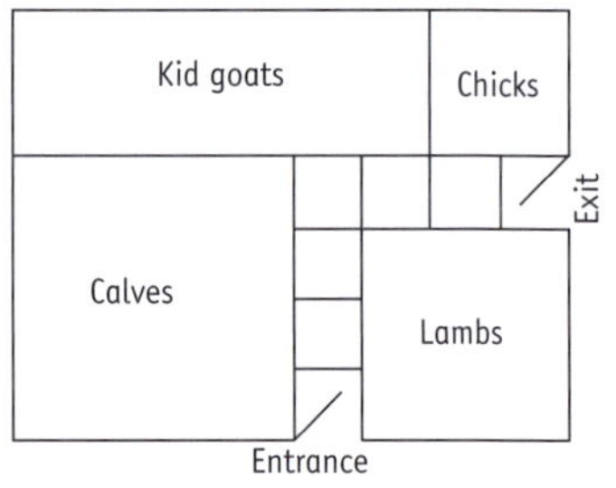

The organiser will need:
- 26 metres of external fencing
- 20 metres of internal fencing
- 7 pavers for the path.

UNIT 20 NUMBER & ALGEBRA

Preparing for Calculator Errors

1	400 × 7 = 2800	500 × 7 = 3500	3150	3213
2	200 × 8 = 1600	300 × 8 = 2400	2000	2072
3	300 × 4 = 1200	400 × 4 = 1600	1400	1488
4	500 × 8 = 4000	600 × 8 = 4800	4400	4272
5	1000 × 7 = 7000	2000 × 7 = 14 000	11 000	10 332

Quick Check

1 36　2 56　3 36　4 35　5 48　6 64　7 81　8 63　9 54　10 21　11 49　12 42　13 45　14 36

Multiplying Money Amounts

1 8 at $19·95 each
8 × $20 = $160
8 × 5c = 40c
$160 – 40c = $159·60

2 4 at $29·75 each
4 × $30 = $120
4 × 25c = $1
$120 – $1 = $119

ANSWERS

TARGETING MATHS HOMEWORK: YEAR 5 © PASCAL PRESS ISBN: 9781925726589

3 12 at $9·90 each
12 × $10 = $120
12 × 10c = $1·20
$120 − $1·20 = $118·80

4 7 at $30·20 each
7 × $30 = $210
7 × 20c = $1·40
$210 + S1·40 = $211·40

5 6 at $120·10 each
6 × $120 = $720
6 × 10c = 60c
$720 + 60c = $720·60

6 8 at $45·55 each
8 × $45 = $360
8 × 55c = $4·40
$360 + $4·40 = $364·40

UNIT 20 STATISTICS & PROBABILITY

Changing Chance

1 1 in 10 or $\frac{1}{10}$
2 1 in 7 or $\frac{1}{7}$
3 1 in 4 or $\frac{1}{4}$
4 Certain

UNIT 20 MEASUREMENT & SPACE

Where is the Treasure?

If the steps are followed correctly, the answer is in Q5 and is (5, 5).

UNIT 20 NAPLAN*-STYLE QUESTIONS

1 $224
2 $\frac{1}{3}$
3 (2, 4)

UNIT 20 PROBLEM OF THE WEEK

Broken Calculator Keys

Many solutions are possible. Examples are:

1 345 × 6
3 4 5 × 1 2 ÷ 2 =

2 768 × 5
7 5 8 × 5 + 5 0 =

3 556 × 6
5 5 5 × 5 + 5 5 7 + 4 =

4 6666 × 9
5 5 5 5 × 9 + 9 9 9 9 =

As above, many key sequences are possible for the next three questions.

5–6 Two examples for 246 × 6 are:
3 5 7 × 5 + 1 3 5 − 4 4 4 =

1 4 5 × 5 + 3 4 7 + 4 0 4 =

7 For 762 × 27, we can use 3 × 3 × 3 = 27 and the key sequence becomes:
8 0 0 − 3 8 = × 3 × 3 × 3 =

UNIT 21 NUMBER & ALGEBRA

Division with Decimal Remainders

1 1·6 or $1 and 60 cents
2 2·25 or 2 m and 25 cm
3 3·25 or $3 and 25 cents
4 1·25 or 1 m and 25 cm
5 3·5 or $3 and 50 cents
6 2·6 or $2 and 60 cents
7 3·8 or 3 m and 80 cm
8 3·5 or $3 and 50 cents

Quick Check – Speedy Division Facts

1 5
2 3
3 8
4 8
5 9
6 9
7 9
8 9
9 7
10 6
11 4
12 7
13 9
14 8
15 12

Number Splitting for Division

1 246
240 + 6
÷ 6
40 + 1
41

2 357
350 + 7
÷ 7
50 + 1
51

3 642
600 + 42
÷ 6
100 + 7
107

4 96
80 + 16
÷ 4
20 + 4
24

5 492
480 + 12
÷ 4
120 + 3
123

6 572
550 + 22
÷ 11
50 + 2
52

UNIT 21 STATISTICS & PROBABILITY

Which Colour?

1 Tick red, yellow and blue. 1 point for each colour selected.
2 Open-ended but similar to “They are all one-third of the whole.” 2 points

UNIT 21 MEASUREMENT & SPACE

Mixed Unit Conversions

1 1·25 L = 1 litre and 250 millilitres or 1250 mL
2 2·36 kg = 2 kilograms and 360 grams or 2360 g
3 3·05 L = 3 litres and 50 millilitres or 3050 mL
4 1·065 kg = 1 kilogram and 65 grams or 1065 g
5 1·7 cm = 1 centimetre and 7 millimetres or 17 mm
6 107 cm = 1 m and 7 cm or 1·07 m
7 1250 g = 1 kg and 250 g or 1·25 kg
8 2705 mL = 2 L and 705 mL or 2·705 L
9 678 cm = 6 m and 78 cm or 6·78 m
10 3005 g = 3 kg and 5 g or 3·005 kg

UNIT 21 NAPLAN*-STYLE QUESTIONS

1 2
2 750 g
3 Second spinner

UNIT 21 PROBLEM OF THE WEEK

Problem of the Week

Open-ended

UNIT 22 NUMBER & ALGEBRA

Division with Remainders

1 6 nets of oranges and 2 oranges left over
2 1$\frac{1}{4}$ blocks of cheese in each tray
3 4 mixing bowls and 800 g left over
4 $5 for each friend and 5c left over
5 3 pieces of ribbon and 10 cm left over

Quick Check

1 3 r 6
2 8 r 1
3 8 r 2
4 8 r 3
5 6 r 6
6 1·25 kg or 1$\frac{1}{4}$ kg
7 1·25 m or 1$\frac{1}{4}$ m
8 1·75 m or 1$\frac{3}{4}$ m
9 20·25 cm or 20$\frac{1}{4}$ cm
10 1·25 L or 1$\frac{1}{4}$ L

Using Number Splitting to Make Estimates of Divisions

1 498
480 + 18
÷ 8
60 + about 2
about 62
Calculator: 62·25

2 364
350 + 14
÷ 7
50 + about 2
about 52
Calculator: 52

3 653
650 + 3
÷ 5
130 + about 1
about 131
Calculator: 130·6

4 513
480 + 33
÷ 6
80 + about 5
about 85
Calculator: 85·5

5 459
420 + 39
÷ 6
70 + about 6
about 76
Calculator: 76·5

6 784
700 + 84
÷ 7
100 + about 12
about 112
Calculator: 112

UNIT 22 STATISTICS & PROBABILITY

Interpreting a Dot Plot

1 16
2 5 seconds
3 19 seconds
4 Half of the runners

UNIT 22 MEASUREMENT & SPACE

Angles of a Triangle

1 48°
2 76°
3 70°
4 55°
5 130°
6 56°

UNIT 22 NAPLAN*-STYLE QUESTIONS

1 $1·20
2 21
3 C

UNIT 22 PROBLEM OF THE WEEK

Chef's Sushi Dinner

Avocado	34
Prawn	68
Beef	126
Chicken	75
Total	**303**

UNIT 23 NUMBER & ALGEBRA

Exploring Multiples and Factors

1 12 = 2 × 6, 6 = 2 × 3; 12 = 3 × 4, 4 = 2 × 2

2 28 = 2 × 14, 14 = 2 × 7; 28 = 7 × 4, 4 = 2 × 2

3 24 = 2 × 12, 12 = 2 × 6, 6 = 2 × 3; 24 = 3 × 8, 8 = 2 × 4, 4 = 2 × 2

4 36 = 2 × 18, 18 = 2 × 9, 9 = 3 × 3; 36 = 3 × 12, 12 = 2 × 6, 6 = 2 × 3

5 42 = 2 × 21, 21 = 3 × 7; 42 = 7 × 6, 6 = 2 × 3

6 54 = 2 × 27, 27 = 3 × 9, 9 = 3 × 3; 54 = 3 × 18, 18 = 3 × 6, 6 = 2 × 3

Quick Check

1 12
2 12
3 20
4 24
5 45
6 1, 2, 4, 8
7 1, 3, 5, 15
8 1, 2, 3, 6, 9, 18
9 1, 2, 3, 4, 6, 8, 12, 24
10 1, 2, 3, 4, 6, 9, 12, 18, 36

Unknowns in Division and Multiplication

1 324 ÷ 6 = 54 and so 324 ÷ 54 = 6
2 456 ÷ 76 = 6 and so 456 ÷ 6 = 76
3 1872 ÷ 24 = 78 and so 1872 ÷ 78 = 24
4 2379 ÷ 39 = 61 and so 2379 ÷ 61 = 39
5 324 ÷ 27 = 12 and so 27 × 12 = 324
6 368 ÷ 46 = 8 and so 46 × 8 = 368
7 665 ÷ 35 = 19 and so 35 × 19 = 665
8 423 ÷ 47 = 9 and so 47 × 9 = 423

UNIT 23 STATISTICS & PROBABILITY

Plotting Plant Growth

1
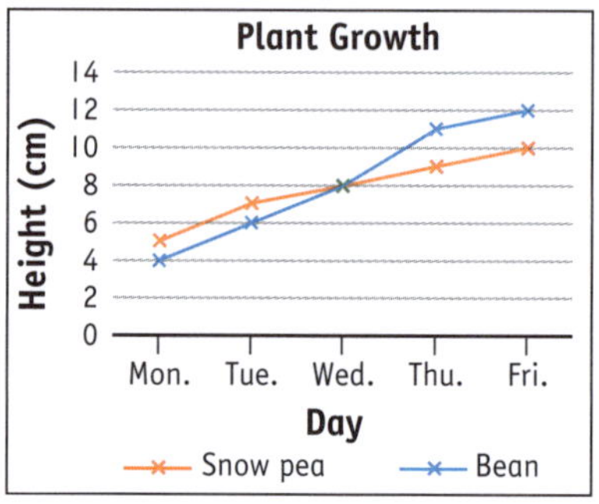

2 Wednesday
3 3 cm
4 Wednesday to Thursday
5 Monday to Tuesday

UNIT 23 MEASUREMENT & SPACE

Using am, pm and 24-hour clock times

1 3:15 am
03:15

2 3:35 pm
15:35

3 12:10 am
00:10

4 12:34 pm
12:34

UNIT 23 NAPLAN*-STYLE QUESTIONS

1 5 2 12 3 Pinocchio

UNIT 23 PROBLEM OF THE WEEK

Time Pressure

A typical solution would be:

Walk dog	25
Ring daughter	10
Packing	20
Taxi ride	35
Total	**90**

This assumes that the Book Taxi item will happen in the same timeslot as Packing. Maria will be at the airport at 8:00 am which is $2\frac{1}{2}$ hours before take-off.

Creating a new timetable is open-ended.

UNIT 24 NUMBER & ALGEBRA

Financial Transactions

1 3 × $1·25 = $3·75
$3·75 + $1·50 = $5·25
$10 – $5·25 = $4·75

2 6 × $0·99 = $5·94
$5·94 + $2·44 = $8·38
$20 – $8·38 = $11·62

3 $65 – $37·50 = $27·50
$27·50 ÷ $2·50 = 11
11 days

4 24 × $9 = $216
$216 + $240 = $456
$456 ÷ 24 = $19

5 28 × $9 = $252
$252 + $240 = $492
$492 ÷ 24 = $20·50

Quick Check

1 $85·50
2 $81·25
3 $63·20
4 $25·75
5 $53·10
6 $2·10
7 S4
8 $8·55
9 $33·40
10 $9

Fractions

1 $\frac{1}{1} = \frac{2}{2} = \frac{3}{3} = \frac{4}{4} = \frac{5}{5} = \frac{6}{6}$

2 $\frac{1}{2} = \frac{2}{4} = \frac{3}{6} = \frac{4}{8} = \frac{5}{10} = \frac{6}{12}$

3 $\frac{1}{3} = \frac{2}{6} = \frac{3}{9} = \frac{4}{12} = \frac{5}{15} = \frac{6}{18}$

4 $\frac{1}{5} = \frac{2}{10} = \frac{3}{15} = \frac{4}{20} = \frac{5}{25} = \frac{6}{30}$

5 $\frac{3}{4} = \frac{6}{8} = \frac{9}{12} = \frac{12}{16} = \frac{15}{20} = \frac{18}{24}$

6 $\frac{2}{3} = \frac{4}{6} = \frac{6}{9} = \frac{8}{12} = \frac{10}{15} = \frac{12}{18}$

7 $\frac{3}{5} = \frac{6}{10} = \frac{9}{15} = \frac{12}{20} = \frac{15}{25} = \frac{18}{30}$

TARGETING MATHS HOMEWORK: YEAR 5 © PASCAL PRESS ISBN: 9781925726589

UNIT 24 STATISTICS & PROBABILITY

Choose a Colour

1 Chart 2
2 The use of colour is confusing – or open-ended.

UNIT 24 MEASUREMENT & SPACE

Translations, Reflections and Rotations

1 Reflection or Flip
2 Translation or Slide
3 Rotation or Turn
4–7 Open-ended

UNIT 24 NAPLAN*-STYLE QUESTIONS

1 $\frac{5}{8}$
2 $118
3 Third diagram

UNIT 24 PROBLEM OF THE WEEK

Charlie the Builder

Yindi should be paid $240.
Anh should be paid $120.
Liam should be paid $320.
Archie should be paid $280.

TERM 3 REVIEW

Review: Unit 17 and Unit 18

1 $0{\cdot}8 = \frac{8}{10}$
2 $0{\cdot}89 = \frac{89}{100}$
3 $0{\cdot}089 = \frac{89}{1000}$
4 $1{\cdot}89 = \frac{189}{100}$
5 $\frac{13}{100} = 0{\cdot}13$
6 $\frac{8}{100} = 0{\cdot}08$
7 $\frac{12}{1000} = 0{\cdot}012$
8 $\frac{176}{100} = 1{\cdot}76$

	The number	is divisible by 2	3	4	5	6	8	9
9	456	✓	✓	✓		✓	✓	
10	6390	✓	✓		✓	✓		✓
11	5472	✓	✓	✓		✓	✓	✓
12	4352	✓		✓			✓	

13 75%
14 20%
15 10%
16 50%
17 1
18 $1\frac{1}{2}$
19 $2\frac{1}{2}$
20 36
21 6
22 Day 1
23 1
24 Open-ended

Review: Unit 19 and Unit 20

1 $\frac{5}{8} + \frac{2}{8} = \frac{7}{8}$
2 $\frac{6}{10} + \frac{1}{10} = \frac{7}{10}$
3 $\frac{3}{6} + \frac{3}{6} = 1$
4 $\frac{6}{10} - \frac{1}{10} = \frac{5}{10}$ or $\frac{1}{2}$
5 $\frac{7}{8} - \frac{2}{8} = \frac{5}{8}$
6 $\frac{5}{6} - \frac{2}{6} = \frac{3}{6}$ or $\frac{1}{2}$
7 Perimeter = 16 m,
Area = 11 m^2
8 Perimeter = 24 m,
Area = 20 m^2
9 High estimate $280
Low Estimate $210
Actual $258·65
10 High estimate $24
Low Estimate $18
Actual $22·50
11 7 × $20 = $140
7 × 50c = $3·50
$140 – $3·50 = $136·50
12 9 × $20 = $180
9 × $1·05 = $9·45
$180 + $9·45 = $189·45
13
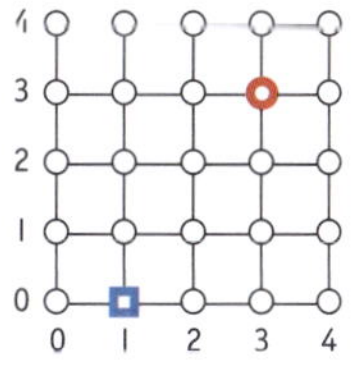

14 (3, 3)
15 (1, 0)

Review: Unit 21 and Unit 22

1 $6·75 or $6 and 75 cents each
2 11·75 m or 11 m and 75 cm
3 175
150 + 25
÷ 5
30 + 5
35
4 91
70 + 21
÷ 7
10 + 3
13
5 108
90 + 18
÷ 9
10 + 2
12
6 567
540 + 27
÷ 9
60 + 3
63
7 1·75 L = 1 L and 750 mL or 1750 mL
8 3·75 kg = 3 kg and 750 g or 3750 g
9 3·65 m = 3 m and 65 cm or 365 cm
10 12 remainder 2
11 $4·25 each
12 552
540 + 12
÷ 9
60 + about 1
about 61
13 325
300 + 25
÷ 3
100 + about 8
about 108
14 45°
15 75°
16 60°

Review: Unit 23 and Unit 24

1
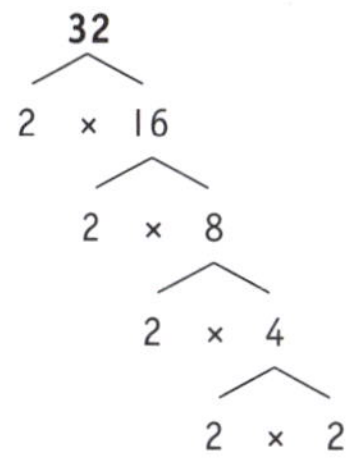

2
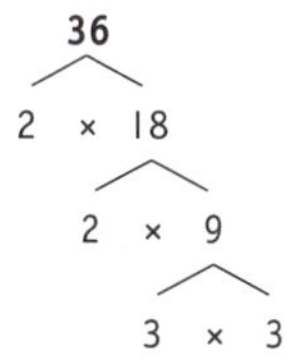

3 Wednesday
4 2 cm
5 12:10 pm
12:10

6 11:50 pm
23:50

7 $50 – $2·90 = $47·10
8 $50 – $17·94 = $32·06
9 $\frac{1}{5} = \frac{2}{10} = \frac{3}{15} = \frac{4}{20} = \frac{5}{25}$
10 $\frac{1}{6} = \frac{2}{12} = \frac{3}{18} = \frac{4}{24} = \frac{5}{30}$
11
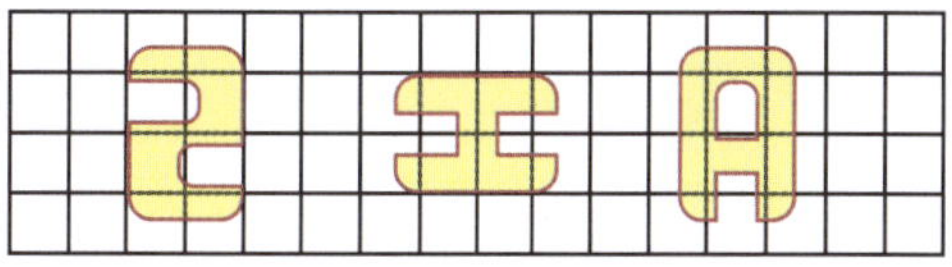

TERM 4

UNIT 25 NUMBER & ALGEBRA

Decimal Place Value

	Fraction	Decimal	Ones	·	Tths	Hdths	Thdths
1	$\frac{9}{1000}$	0·009	0	·	0	0	9
2	$\frac{37}{1000}$	0·037	0	·	0	3	7
3	$\frac{1456}{1000}$	1·456	1	·	4	5	6
4	$\frac{375}{100}$	3·75	3	·	7	5	
5	$\frac{456}{100}$	4·56	4	·	5	6	

6 2·608 is 2 ones, 6 tenths, 0 hundredths and 8 thousandths.
7 2·685 is 2 ones, 6 tenths, 8 hundredths and 5 thousandths.
8 3·009 is 3 ones, 0 tenths, 0 hundredths and 9 thousandths.
9 3·079 is 3 ones, 0 tenths, 7 hundredths and 9 thousandths.
10 3·76 is 3 ones, 7 tenths, 6 hundredths and 0 thousandths.

Quick Check

1 6
2 3
3 1
4 0
5 5
6 6
7 5
8 6
9 5
10 0

Factors and Multiples

	Numbers to use	Fact.	Mult.	Fact.	Mult.	Fact.	Mult.
1	2, 3, 5, 6, 15, 30	3	6	2	30	5	15
2	2, 3, 6, 9, 12, 18	6	18	2	12	3	9
3	2, 4, 8, 10, 12, 20	10	20	4	12	2	8
4	2, 5, 10, 20, 50, 100	2	50	5	20	10	100
5	6, 9, 18, 27, 36, 54	18	36	6	54	9	27

ANSWERS

UNIT 25 STATISTICS & PROBABILITY

Line Graphs

1 400 km
2 7 hours
3 Between hours 3 and 4
4 Open-ended but they might have stopped for lunch.
5 Open-ended but there could have been too much traffic.

UNIT 25 MEASUREMENT & SPACE

Angles of a Quadrilateral

1 60° + 90° + 90° + 120° = 360°
2 105° + 75° + 75° + 105° = 360°
3 105° + 105° + 75° + 75° = 360°
4 80° + 60° + 130° + 90° = 360°
5 The angles always add to 360°.

UNIT 25 NAPLAN*-STYLE QUESTIONS

1 36
2 B
3 The Fun Run was uphill for the first half of the race.

UNIT 25 PROBLEM OF THE WEEK

Factors and Multiples on a Hexagon

Sample answer

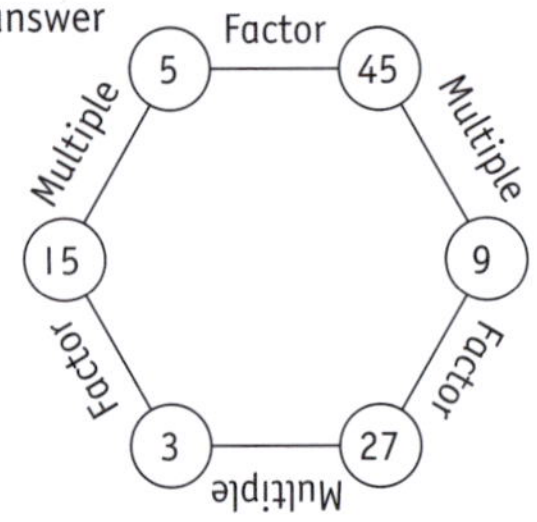

UNIT 26 NUMBER & ALGEBRA

Comparing Fractions

1
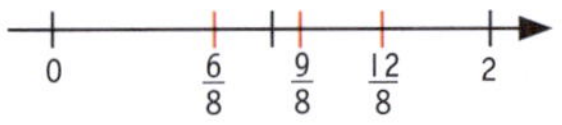

2
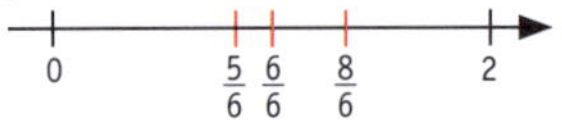

3
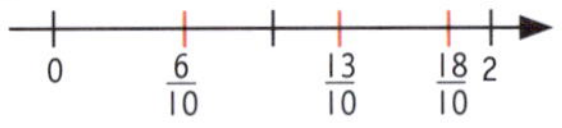

4
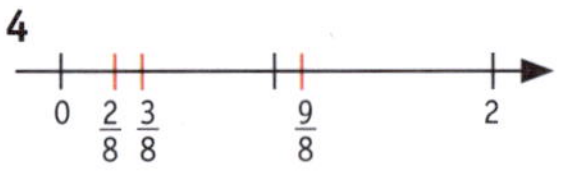

5
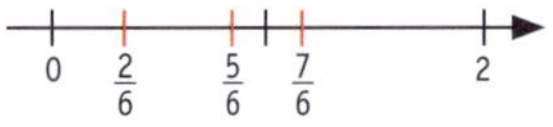

6
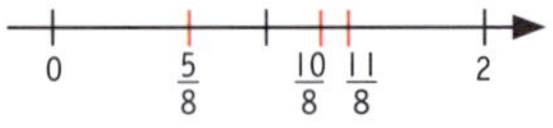

Quick Check

1 $\frac{3}{4}$　3 $\frac{1}{4}$　5 $\frac{1}{2}$　7 $\frac{1}{10}$　9 $\frac{1}{12}$
2 $\frac{5}{8}$　4 $1\frac{1}{4}$　6 $\frac{1}{3}$　8 $\frac{3}{16}$　10 $\frac{1}{32}$

Decimals to Fractions

1 \$35　3 \$40　5 \$75　7 \$25　9 \$24
2 \$25　4 50c　6 \$6·25　8 \$135　10 \$2500

UNIT 26 STATISTICS & PROBABILITY

Design a Spinner

1
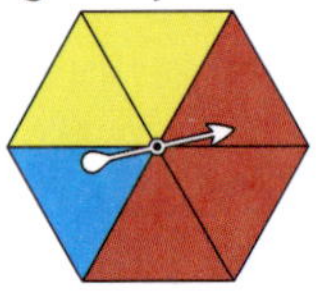
2
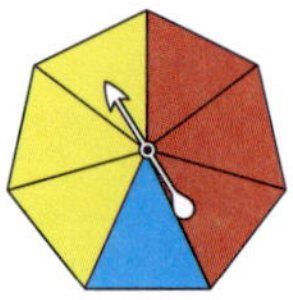
3
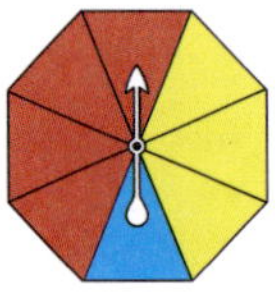

UNIT 26 MEASUREMENT & SPACE

Transformations

1
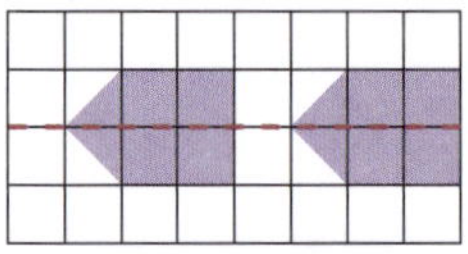
2
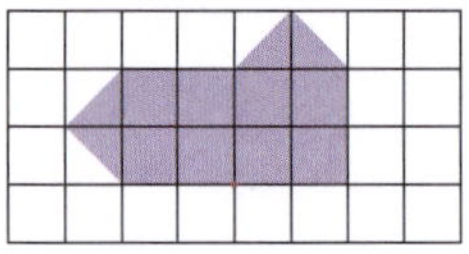
3
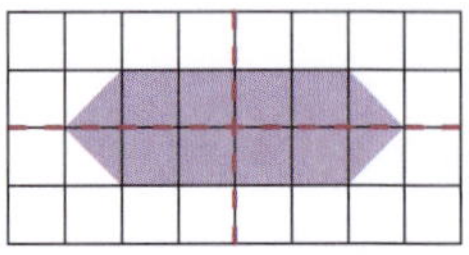
4 As shown on the diagrams.
5 3
6 2
7
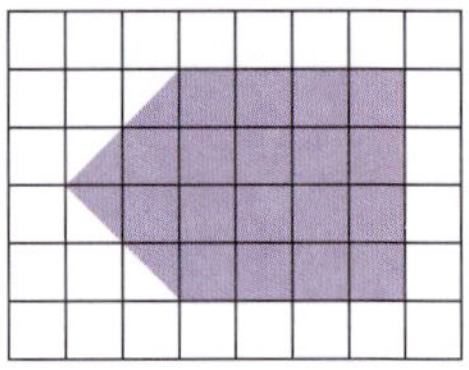

UNIT 26 NAPLAN*-STYLE QUESTIONS

1 $\frac{2}{12}$
2
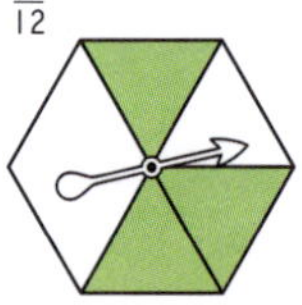
3 Reflection

UNIT 26 PROBLEM OF THE WEEK

Race to 1

Will could win with $\frac{1}{4}$ or $\frac{2}{6}$.

Asha could win with $\frac{6}{8}$ or $\frac{4}{6}$.

Ravi could win with $\frac{3}{8}$, $\frac{2}{6}$ or $\frac{1}{2}$ and so has the best chance of winning in Round 4.

UNIT 27 NUMBER & ALGEBRA

Adding Fractions

1 $2\frac{1}{4}$　3 $2\frac{3}{4}$　5 $2\frac{1}{6}$　7 $2\frac{2}{5}$
2 $2\frac{3}{8}$　4 $2\frac{2}{3}$　6 $2\frac{1}{4}$

Quick Check

1 $1\frac{1}{2}$　3 $1\frac{1}{2}$　5 $1\frac{1}{6}$　7 $1\frac{2}{5}$　9 $1\frac{5}{6}$
2 $1\frac{3}{10}$　4 $1\frac{1}{4}$　6 $1\frac{1}{3}$　8 $1\frac{1}{4}$　10 2

Multiplication is Commutative

Note: The order of the multiplications is open-ended, but the result should be the same.

1 4 × 7 = 28, 28 × 3 = 84, 84 × 2 = 168
2 5 × 4 = 20, 20 × 6 = 120, 120 × 3 = 360
3 2 × 5 = 10, 10 × 7 = 70, 70 × 4 = 280
4 8 × 5 = 40, 40 × 6 = 240, 240 × 3 = 720
5 9 × 5 = 45, 45 × 2 = 90, 90 × 3 = 270

UNIT 27 STATISTICS & PROBABILITY

Borrowed Books

1
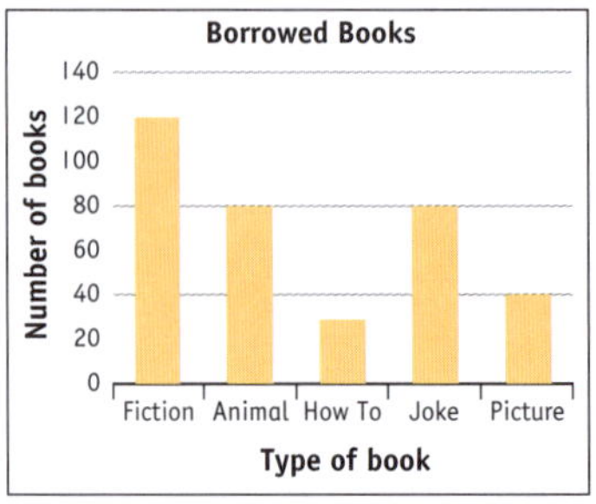

2 80
3 350
4 Picture + Animal or Joke
5 40

ANSWERS

TARGETING MATHS HOMEWORK: YEAR 5 © PASCAL PRESS ISBN: 9781925726589

UNIT 27 MEASUREMENT & SPACE

Designing a City Skyline

Open-ended.

UNIT 27 NAPLAN*-STYLE QUESTIONS

1 $\frac{3}{4}$ 2 30 cm 3 46 bread rolls

UNIT 27 PROBLEM OF THE WEEK

Estimate and Check

While the estimates are open-ended, Challenge 1 measurements should be close to: E – 77 mm, J – 47 mm, L – 47 mm, S – 83 mm. Challenge 2 is open-ended.

UNIT 28 NUMBER & ALGEBRA

Division

1 456 ÷ 88
÷ 8
57 ÷ 11
5 r 2

2 372 ÷ 12
÷ 6
62 ÷ 2
31

3 584 ÷ 72
÷ 8
73 ÷ 9
8 r 1

4 488 ÷ 32
÷ 8
61 ÷ 4
15 r 1

5 574 ÷ 35
÷ 7
82 ÷ 5
16 r 2

6 364 ÷ 28
÷ 7
52 ÷ 4
13

Quick Check

1 No 2 Yes 3 No 4 No 5 No 6 Yes 7 Yes 8 Yes 9 Yes 10 No

Multiplication

The order of the factors is open-ended, but the products are:
1 432 2 210 3 216 4 288 5 384 6 196

UNIT 28 STATISTICS & PROBABILITY

Reading a Chart

1 Monday 2 Wednesday 3 Thursday 4 120 5 70

UNIT 28 MEASUREMENT & SPACE

Area and Perimeter

1 Open-ended
2 Open-ended
3 Shape D
4 Shape B

UNIT 28 NAPLAN*-STYLE QUESTIONS

1 $24·50 2 183, 263 3 First shape

UNIT 28 PROBLEM OF THE WEEK

Making 2-digit Numbers

Challenge 1
Asha should make 53 × 72 = 3816.
Ravi should make 25 × 37 = 925.

Challenge 2
Asha should make 75 × 83 = 6225.
Ravi should make 37 × 58 = 2146.

Challenge 3
Asha should make 532 × 7 = 3724 and Ravi should make 357 × 2 = 714.
Asha should make 753 × 8 = 6024 and Ravi should make 578 × 3 = 1734.

UNIT 29 NUMBER & ALGEBRA

Decimal Multiplication

1

	Rolls	Donuts	Muffins	Sausage Rolls
1	$1·15	$2·35	$1·85	$3·35
2	$2·30	$4·70	$3·70	$6·70
3	$3·45	$7·05	$5·55	$10·05
4	$4·60	$9·40	$7·40	$13·40
5	$5·75	$11·75	$9·25	$16·75

2 $24·70 3 $59·20 4 $63·00 5 $104·40 6 $74·10

Quick Check

1 $9·20 2 $12·95 3 $23·50 4 $36·85 5 $21·15 6 $8·70 7 $87·00 8 $29·40 9 $67·00 10 $74·00

Lowest Common Multiples

1 6 × 8
3 × 2 × 4 = 24

2 15 × 21
5 × 3 × 7 = 105

3 12 × 16
3 × 4 × 4 = 48

4 18 × 24
3 × 6 × 4 = 72

5 6 × 14
3 × 2 × 7 = 42

UNIT 29 STATISTICS & PROBABILITY

Morning Temperatures

1 6°
2 5°
3 On Tuesday between 7 am and 8 am
4 3°
5 Open-ended

UNIT 29 MEASUREMENT & SPACE

Time Zones

1 11:30 am, same day
2 9:00 pm, previous day
3 6:00 pm, previous day
4 10:00 pm, same day
5 2:00 pm, same day
6 2:15 am, next day
7 1:00 am, same day
8 7:45 pm, previous day

UNIT 29 NAPLAN*-STYLE QUESTIONS

1 132, 1344, 252, 372
2 6 × $4 – 6 × 10c
3 6

UNIT 29 PROBLEM OF THE WEEK

Morning Tea Breaks

1 1 sausage roll and 2 fruit skewers
2 2 cheese skewers and 2 ham roll-ups
3 2 fruit skewers and 3 sausage rolls
4 4 muffins and 2 ham roll-ups

	Sausage Roll	Cheese Skewer	Fruit Skewer	Muffin	Ham Roll-up
Price	$2·50	$3·10	$2·75	$3·30	$3·90
1	$2·50	$3·10	$2·75	$3·30	$3·90
2	$5·00	$6·20	$5·50	$6·60	$7·80
3	$7·50	$9·30	$8·25	$9·90	$11·70
4	$10·00	$12·40	$11·00	$13·20	$15·60

UNIT 30 NUMBER & ALGEBRA

Multiplication and Division

1 6 × 4 = 24 2 24 ÷ 6 = 4 3 24 ÷ 4 = 6 4 4 × 6 = 24 5 4 × 8 = 32 6 32 ÷ 8 = 4 7 8 × 4 = 32 8 32 ÷ 4 = 8

Quick Check

1 3 × 5 = 15, 5 × 3 = 15, 15 ÷ 3 = 5, 15 ÷ 5 = 3
2 3 × 7 = 21, 7 × 3 = 21, 21 ÷ 3 = 7, 21 ÷ 7 = 3
3 6 × 7 = 42, 7 × 6 = 42, 42 ÷ 6 = 7, 42 ÷ 7 = 6
4 6 × 9 = 54, 9 × 6 = 54, 54 ÷ 6 = 9, 54 ÷ 9 = 6
5 7 × 8 = 56, 8 × 7 = 56, 56 ÷ 7 = 8, 56 ÷ 8 = 7
6 3 × 9 = 27, 9 × 3 = 27, 27 ÷ 3 = 9, 27 ÷ 9 = 3
7 4 × 7 = 28, 7 × 4 = 28, 28 ÷ 4 = 7, 28 ÷ 7 = 4
8 7 × 9 = 63, 9 × 7 = 63, 63 ÷ 7 = 9, 63 ÷ 9 = 7
9 3 × 11 = 33, 11 × 3 = 33, 33 ÷ 3 = 11, 33 ÷ 11 = 3
10 3 × 31 = 93, 31 × 3 = 93, 93 ÷ 3 = 31, 93 ÷ 31 = 3

Multiplication of Money Amounts

1

	Dollars	Cents		
	3	60	5	
× 8	24	480	40	= $29·20

2

	Dollars	Cents		
	2	50	5	
× 7	14	350	35	= $17·85

3

	Dollars	Cents		
	4	60	3	
× 5	20	300	15	= $23·15

4

	Dollars	Cents		
	4	80	9	
× 3	12	240	27	= $14·67

5

	Dollars	Cents		
	3	30	5	
× 4	12	120	20	= $13·40

6

	Dollars	Cents		
	6	70	2	
× 8	48	560	16	= $53·76

UNIT 30 STATISTICS & PROBABILITY

In the Bag

1

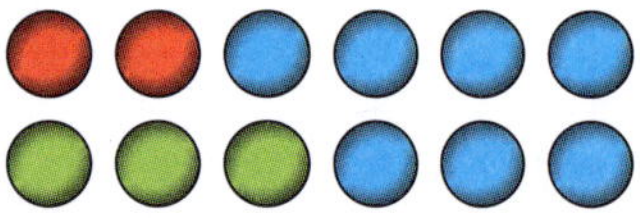

2 7 in 12 chance 3 2 green counters 4 1 in 7 chance

UNIT 30 MEASUREMENT & SPACE

3D Nets

1

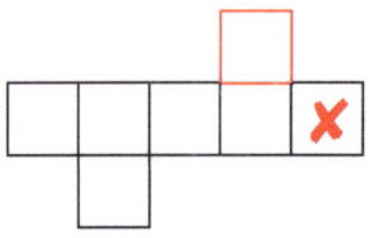

2

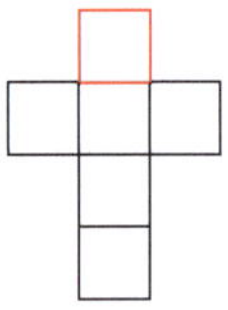

3

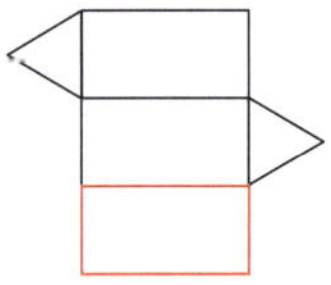

4

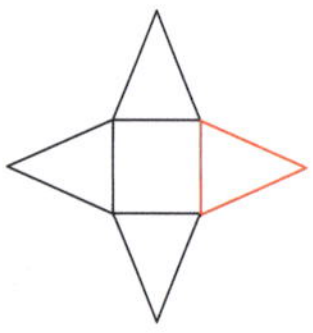

5

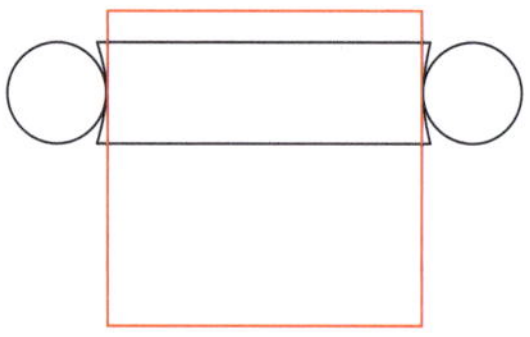

6

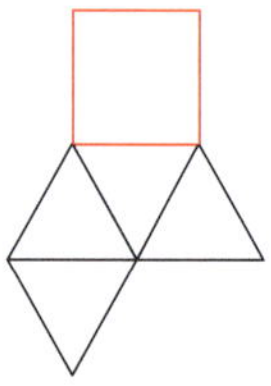

UNIT 30 NAPLAN*-STYLE QUESTIONS

1 Third option: The greengrocer has to share his 48 oranges equally between 8 nets. How many oranges should he put into each net?
2 5 in 6
3 Rectangular prism

UNIT 30 PROBLEM OF THE WEEK

A Puzzle Kit

Will can make 1 cylinder, 1 triangular pyramid, 1 cube, 1 rectangular prism and 1 triangular prism.
Extra challenge: Will would have 1 square and 4 rectangles left over or 5 squares.

UNIT 31 NUMBER & ALGEBRA

Percentages

1 37%
2 64%
3 48%
4 60%
5 63%
6 48%
7 Yellow 21%, Green 30%, Blue 49%
8 Yellow 33%, Green 35%, Blue 32%
9 Yellow 32%, Green 47%, Blue 21%

Quick Check

1 30% 2 65% 3 15% 4 37% 5 76% 6 34% 7 43% 8 42% 9 53% 10 85%

Fraction Word Problems

1 $1\frac{1}{4}$ h 2 $1\frac{3}{4}$ m 3 $1\frac{1}{4}$ kg 4 $3\frac{1}{4}$ L 5 50 km 6 375 km

UNIT 31 STATISTICS & PROBABILITY

On the Road

Open-ended, but answers should be similar to:
1 5 2 10 3 2 4 10 5 0

UNIT 31 MEASUREMENT & SPACE

Angles

Open-ended, except for Q3 where the angles add to 360°.

UNIT 31 NAPLAN*-STYLE QUESTIONS

1 $43 2 D 3 20

UNIT 31 PROBLEM OF THE WEEK

The Calculator Kids

1 The sequence generated by 4 + 3 = = is:
4, 7, **10**, 13, 16, 19, **22**, 25, 28, 31, **34**, 37
2 The sequence generated by 2 + 4 = = is
2, 6, **10**, 14, 18, **22**, 26, 30, **34**, 38, 42, 46

An interesting challenge would be to explain why the common numbers are 12 apart each time!

Had another go

3 The sequence generated by 3 + 4 = = is
3, 7, 11, 15, 19, 23, 27, 31, 35, 39, 43, 47
4 The sequence generated by 2 + 6 = = is
2, 8, 14, 20, 26, 32, 38, 44, 50, 56, 62, 68

One sequence is made up of all odd numbers after the start while the second sequence is made up of only even numbers. These sequences will never have a common entry.

The Challenge

If the start number is the same for both sequences, then whatever the second numbers added on each time, the first common entry will happen when the lowest common multiple of the two second numbers is reached.

Making sequences that are always odd or always even will be a way to make them never have a common entry.

UNIT 32 NUMBER & ALGEBRA

Finding Unknowns

1 7 × ? + 5 − **5** = 61 − **5**
7 × ? = **56**
7 × **8** = **56**

2 4 × ? + 3 − **3** = 67 − **3**
4 × ? = **64**
4 × **16** = **64**

3 6 × ? − 5 + **5** = 37 + **5**
6 × ? = **42**
6 × **7** = **42**

4 ? × 6 + 3 − **3** = 75 − **3**
? × 6 = **72**
12 × 6 = **72**

5 ? × 5 + 6 − **6** = 126 − **6**
? × 5 = **120**
24 × 5 = **120**

6 ? × 8 − 5 + **5** = 51 + **5**
? × 8 = **56**
7 × 8 = **56**

TARGETING MATHS HOMEWORK: YEAR 5 © PASCAL PRESS ISBN: 9781925726589

Quick Check

1 12	3 50	5 21	7 3	9 175
2 3	4 8	6 8	8 60	10 220

Unknowns in Money Problems

1 $? × 6 = $72
$? = $72 ÷ 6
$? = $12

2 $? × 8 = $96
$? = $96 ÷ 8
$? = $12

3 $? × 3 = $18·15
$? = $18·15 ÷ 3
$? = $6·05

4 $? × 7 = $84·70
$? = $84·70 ÷ 7
$? = $12·10

5 $? × 4 = $4·44
$? = $4·44 ÷ 4
$? = $1·11

6 $? × 9 = $108·90
$? = $108·90 ÷ 9
$? = $12·10

UNIT 32 STATISTICS & PROBABILITY

Family Movie Night

Open-ended

UNIT 32 MEASUREMENT & SPACE

Hidden Picture

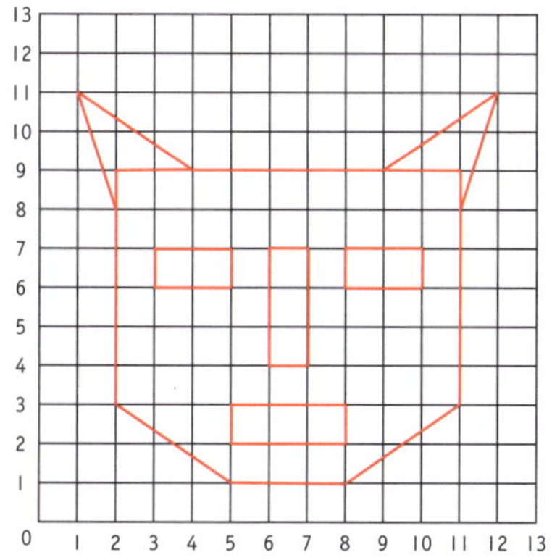

UNIT 32 NAPLAN*-STYLE QUESTIONS

1 $7·40
2 (3, 2)
3 8 × 16 = 128

UNIT 32 PROBLEM OF THE WEEK

The Fundraiser

1 15 trays of sausages, 10 bags of rolls
2 $432
3 15 trays of sausages @ $7·95 = $119·25
10 trays of rolls @ $10·40 = $104
Profit = $432 − 223·25 = $208·75
4 Open-ended, but probably, "Charge more for the hot dogs!"

TERM 4 REVIEW

Review: Unit 25 and Unit 26

1 $\frac{11}{1000}$ is 0 ones, 0 tenths, 1 hundredths and 1 thousandths

2 $\frac{109}{1000}$ is 0 ones, 1 tenths, 0 hundredths and 9 thousandths

3 $\frac{19}{10}$ is 1 ones, 9 tenths, 0 hundredths and 0 thousandths

4 9:00
5 250 km
6 10:00 and 11:00
7 11:00
8 Angle A is 135°
Angle B is 45°
Angle C is 90°
Angle D is 90°
9 $\frac{15}{8}, \frac{6}{8}, \frac{12}{8}$
10 $\frac{4}{10}, \frac{5}{10}, \frac{13}{10}$
11 $9·50
12 $52·50
13 $12·50
14 $37·50
15

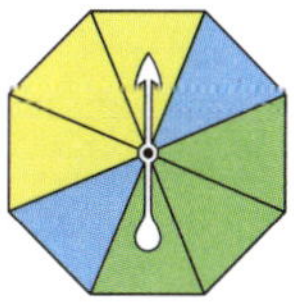

Review: Unit 27 and Unit 28

1 $\boxed{\frac{3}{4}} + \frac{1}{5} + \boxed{\frac{2}{8}} = 1\frac{1}{5}$

2 $\boxed{\frac{3}{5}} + \frac{1}{2} + \boxed{\frac{4}{10}} = 1\frac{1}{2}$

3 $\boxed{\frac{1}{3} + \frac{4}{6}} + \frac{3}{5} = 1\frac{3}{5}$

4 $\boxed{\frac{5}{8}} + \boxed{\frac{1}{4}} + \frac{1}{6} + \boxed{\frac{1}{8}} = 1\frac{1}{6}$

5 5 × 6 = 30, 30 × 4 = 120,
120 × 3 = 360
6 6 × 5 = 30, 30 × 7 = 210,
210 × 2 = 420
7 60
8 180
9 50
10 365 ÷ 15
÷ 5
73 ÷ 3
24 r 1
11 432 ÷ 18
÷ 6
72 ÷ 3
24 r 0
12 352 ÷ 16
÷ 8
44 ÷ 2
22 r 0
13 18 × 16
9 × 2 × 8 × 2
9 × 8 × 2 × 2 = 288
14 24 × 15
8 × 3 × 5 × 3
8 × 5 × 3 × 3 = 360
15 blue and yellow
16 blue and orange

Review: Unit 29 and Unit 30

1

	Pencils	Pens	Notebooks	Erasers
1	$1·35	$1·95	$2·60	$0·89
2	$2·70	$3·90	$5·20	$1·78
3	$4·05	$5·85	$7·80	$2·67
4	$5·40	$7·80	$10·40	$3·56

2 $13·50
3 $39
4 9 12
3 × 3 × 4 = 36
5 5:00 pm, same day
6 48 bananas
7 9 bananas
8 7 bunches
9 $2·65 × 6

	Dollars	Cents	
	2	60	5
× 6	12	360	30

= $15·90

10 $3·36 × 7

	Dollars	Cents	
	3	30	6
× 7	21	210	42

= $23·52

11 $8·43 × 3

	Dollars	Cents	
	8	40	3
× 3	24	120	9

= $25·29

12 $6·45 × 8

	Dollars	Cents	
	6	40	5
× 8	48	320	40

= $51·60

ANSWERS

Review: Unit 31 and Unit 32

1 63%
2 76%
3 65%
4 43%
5 16%
6 39%
7 1 h 30 min
8 130 mm
9 135°
10 45°
11 225°
12 8 × ? + 6 − **6** = 54 − **6**
8 × ? = **48**
8 × **6** = **48**
13 ? × 6 + 5 − **5** = 47 − **5**
? × 6 = **42**
7 × 6 = **42**
14 ? × 6 = **$96**
? = **$96 ÷ 6**
? = **$16**
15 ? × 8 = **$96·80**
? = **$96·80 ÷ 8**
? = **$12·10**
16

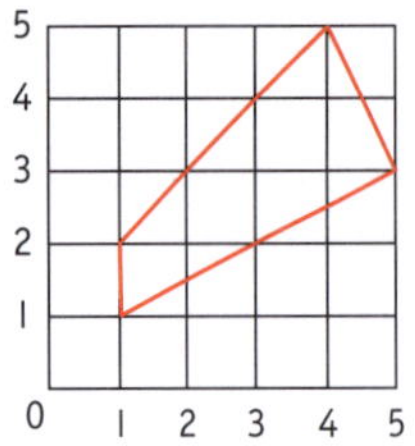

17 1
18 2
19 quadrilateral

TARGETING MATHS HOMEWORK: YEAR 5 © PASCAL PRESS ISBN: 9781925726589